The Quantum Method

The Quantum Method

The science of productivity

Mike Tranter PhD

Copyright © 2022 Mike Tranter The English Scientist

First paperback edition November 2022

All rights reserved. No part of this book may be reproduced or used in any manner without written permission of the copyright owner except for the use of quotations in a book review. For more information, contact: neurorevolution@protonmail.com.

Cover design by Nouman Sarwar

Edited by S. J. Rose & Christina Houen

Author photograph copyright Art Kroetz 2021

ISBN 978-1-737-1026-6-3 (paperback)

www.aNeuroRevolution.com

1 2 3 4 5 6 7 8 9 10

All rights reserved.

DEDICATION

This book is dedicated to every reader who wants to change their life. For anyone who wants to find improvement and aim higher, even if it is only in the smallest way. For anyone willing to put aside their fear, doubt, and self–criticism, to reach a level they always knew existed, but never dared to reach before, until today.

TABLE OF CONTENTS

Page	Chapter
	PREFACE
1	THE SCIENCE BEHIND MOTIVATION AND DRIVE
10	CREATING A BASELINE FOR YOURSELF
25	EASY, MEDIUM, OR HARD?
37	UNDERSTANDING YOUR TIME AND EFFICIENCY
54	LEARN WHAT WORKS BEST FOR YOU
68	ZONAL WORKING
75	ANYTIME, ANYWHERE
83	GOALS AND REWARDS
93	MAKING A LASTING CHANGE
101	PUTTING IT INTO PRACTICE
108	ABOUT THE AUTHOR
109	BOOKS BY THE AUTHOR
110	REFERENCES

PREFACE

Each of us has untapped power and the potential to be great. Unfortunately, for most people, that drive for greatness lies dormant and unused until eventually, we tell ourselves that it doesn't exist anymore. We are too busy right now. We would do it if we were younger and more courageous. Now isn't the right time, maybe next year. The truth is, today is your day, right now. **Today!**

We all have busy lives, and it can be challenging to fit everything into your day. Of course, we still want to feel we are on the path to something greater — something down the line, towards a future we envision for ourselves. The result of our hectic lifestyles is that for each of these dreams and ambitions, while important to us, we do not have the time to pursue them. So, what happens? We settle for second best and give up on the dreams we used to have. They fade into our past and are never brought into our reality.

For many people, that is OK. As human beings, we are resourceful, and so we learn to adapt and realign our dreams to our current circumstances, especially when new events in our lives take priority. Want to save up to buy a house but just had a baby? Well, of course, saving now goes out the window when you have another hungry (and adorable) mouth to feed. Want to start on a new career path? Doing it now wouldn't be appropriate, you might get that promotion you have been waiting for. Want to start a business? Well, of course that has to wait, you barely have time for your friends and family now — taking on something this big just isn't possible.

What if there is a way to have both? What if there was a way to keep a hold of your dreams and never let them fade into the

background, while you balance everything in your schedule? That is why we are here.

Why even bother to upset the status quo and try to change your current circumstances? One survey found that as many as 70% of workers are not emotionally engaged in their job.[1] This is one of the reasons why it is so crucial to invest your energy in your dreams. Occasionally those dreams may involve working in the job that you already have, in which case, you can apply the techniques and concepts of this book to your professional life and elevate your level above that of anyone else around you. For the remaining 70%, this book should serve as a reminder to spend your energy achieving your dreams. As Steve Jobs, co–founder of Apple once famously said, "the only way to do great work is to love what you do".

The concept of this book

We all have the same 24 hours in the day. The issue is, that those 24 hours are filled with everything else. We need to sleep, work, commute and travel, raise our children, eat, read, socialise, and a million other things. If that is all you want, then great, there is absolutely nothing wrong with that. But, if you're going to achieve more and build your own path to the future you want, then it is time to take a candid look at your schedule and an even more honest look at yourself. In this book, we will identify ways that you can increase your productivity regardless of your busy schedule, and map out exactly how you will surpass your limits to take a firm hold of your dreams.

Essentially, *The Quantum Method* breaks down your day into smaller fragments. Some can be as small as five minutes and others a few hours. It teaches you to question yourself in a way that will allow you to unlock your potential. You will learn precisely how to perform at your best, and, importantly, how to overcome those times when you don't. You will learn to fill your time with an appropriate level of work tailored to your personal schedule and energy level. The concept behind *The Quantum Method* is simple; by learning how you schedule your life, you can optimise your efforts to increase your efficiency. Small gains you earn throughout the week will always work towards the bigger dream. When you look back ten years from now, it won't be some grand stroke of luck that got you there, it will be the culmination of small, consistent, and disciplined efforts that all contributed to your success.

The methods and techniques used within this book are based on scientific research and the most current theories in psychology. As a neuroscientist, I feel that a lot of books and articles that promise improvements are not based on tried and tested methods which have been scrutinised by the scientific community to ensure their quality. Instead, they are based on general ideas and fashionable statements that are not necessarily rooted in scientific thinking. I wanted to write something different.

For each chapter, I include explanations regarding the science and psychology behind why these methods work, how we can use them to our advantage, and what factors may influence their outcomes. The result we are after, in each case, is increased productivity that helps you work towards your dream and goals. The scientific literature used as a foundation for the techniques within these chapters can be found in the reference section at the back of the book.

While it is true that *The Quantum Method* has been designed to fit into anybody's life and schedule, it will be up to you to determine how much time you can put into working on your goals. If you only have ten minutes spare each day, we will work together to understand how to fill those ten minutes with as much energy and enthusiasm as possible. If you have three hours, we will customise a structure for those hours to give you the best chance of success.

A crucial element that is often underestimated is that efficiency and a deeper understanding of yourself are more impactful than you may ever have realised. After all, a real and sustainable change starts with looking inward, so we can understand what impedes us from making a permanent change. Whenever I discuss these methods with people who are trying to improve their productivity, they are often surprised at the level of detail that is required to restructure their day to boost performance. This is especially true when a person already has an abundance of responsibilities and commitments in their life. In those instances, it will require a forensic look into the timings of each day, because it will not be immediately obvious where the spare time to peruse your goals is to be found. At first, the adjustment to your schedule can be a shock. It is never easy when we decide to make a change in our lives, whatever that change might be, and it is never easy to stick with it for the long term. But, improving your productivity and achieving more than anyone around you will be worth it in the end.

How to use this book

The initial part of this book explores some of the fundamental concepts around goal setting and motivation. By having an appreciation for some of the psychology behind these processes, we can learn to identify them when they occur throughout our own journey. They can also serve as a basis for scientific explanation of why many of the techniques within this book will work for you, because there have been decades of research into understanding exactly how to get the most out of your time so you can have the life you want.

The remainder of the book is broken down into ten chapters, with each one describing a particular technique or style of analysis to improve your productivity. Each chapter also begins with a motivational phrase that vividly expresses the ethos of the chapter. Naturally, the chapters are designed to be read in sequence, meaning that the most significant benefit will come to those who go through each chapter, building on the previous techniques from the earlier chapters. But it can work another way too. It can work well if you handpick the necessary elements that work for your life. In this way, the book chapters are brief enough for you to quickly implement some of the techniques, and you can absorb the fundamentals of what the system is about and get straight to work. I hope this helps you to read through the chapters and find what is most appealing and valuable for you as an individual.

We all enjoy talking about what we want to do, or what we will do *one day*. But we rarely put a plan into practice. We don't dissect our life and see how we need to work towards that goal. We don't examine our dream to see how it can be broken down into manageable amounts to be worked on every day. We typically see it only as an end product, something we can simply wish into existence. We rarely analyse our unique working styles and habits to see how we can use our personal qualities to our advantage. An old adage that suits this book perfectly is that you miss 100% of the opportunities you never take. So, in order to create that opportunity to bring your dream into a reality, you need to go out and get it, using everything you have!

You don't need to be extraordinary to do extraordinary things. You just need to be honest with yourself, be consistent, and never be afraid to use whatever it is that makes you great. In this book, you will learn to do it all.

1

THE SCIENCE BEHIND MOTIVATION AND DRIVE

This is not my final phase. This is not who I am always going to be. I am not finished. I am just beginning...

While many books and courses provide excellent resources to improve productivity, there is a consistent issue with their approach. They are aimed at the general reader and as a consequence, they can offer only broad advice — a one–type–suits–all approach. *The Quantum Method* works because it was founded on the premise that the greatest benefits will come from a system designed to work for you as an individual. What may be suitable for one person will not necessarily show the same results for another.

An example of this would be choosing what time in the day you should work on your tasks, or better still, the time when you should complete your most intense workload. Waking up early is a great example of this. Are you someone who prefers waking up early and ensuring that your work is completed before the rest of the world has started its day? Or are you someone who needs a slower start so your energy can pick up in the evening? Understanding the answer to this question is not necessarily about permitting yourself to sleep past your alarm because you prefer to wake up later; instead, it is about identifying when you should prioritise the most intensive tasks on your list to get the most out of your time.

We want to learn how to identify your strengths and preferred working habits and formulate a new schedule to fit your daily life. One approach to this is to explore the areas of your life that are contributing to your productivity in order to design a program to maximise your achievements. This will require an initial stage of critical thinking and planning. A journal or notebook is recommended as a place where you can write out your schedule and design your personalised system. This will help to organise your new routine and serve as a safe storage location where your productive insights and written goals can be logged. You can also return to it at any time and reflect on your progress whenever you need to.

There is an accompanying journal available for this book that is designed to do precisely that and is already structured in a way to complement this book as we go through each chapter. But, if you don't use *The Quantum Method Journal*, you can use any notepad you have nearby. It will be instrumental in writing out the answers to the questions set out below and vital for when you look back at the notes in the coming months to keep track of your plan and vision for the future. It is easy to get lost in some of the more minor details of life and forget about the bigger dream. These notes will help you to hold on to that dream and help you navigate your life in times of stress and doubt. A calendar will also be useful for this reason, and is worth considering. This can be a physical calendar that is placed on a wall in your house, or an electronic copy on a phone or computer. Electronic calendars are becoming more popular because they can be readily accessed whenever you need them.

A research study looking at groups of people who planned their schedule compared with those who didn't, noted that planning can help a person to complete their tasks faster and more accurately.[2] What is even better is that in this study, the group who actively planned their tasks had a greater overall completion rate, which helped them to improve their daily productivity. Calendars work particularly well with to–do lists because they assist our prospective memory. Prospective memory is what we use to remember items we need to do and the correct time to do them, such as appointments or meetings throughout the day. If you have never been the type of person to use calendars or write down your to–do lists, instead relying on memory, then this could be a great opportunity to try something new.

The reason that you will achieve your dream

Where does motivation come from? Is it something we can train and develop, or is it innate within ourselves? What makes some people highly motivated and others much less so? Psychologists have had a keen interest in understanding the source of motivation since the 1950s when there was an incentive to improve motivation amongst industrial workers to increase performance and maximise profits.[3] In fact, even as early as the 1930s, scientists could already see the need to understand the effects of motivation on performance.[4]

Initially, there was an academic debate in the scientific community as to whether motivation was a subconscious or conscious feature of the mind; in other words, whether a person could control their level of motivation, or if it derived from a part of the mind we could never directly influence. The subconscious theory would essentially mean that some individuals are born highly motivated, while others are more likely to fail in their attempts to consistently work towards their goals. We now know, of course, that motivation is highly influenced by our current mindset. Nearly a century of psychology research has provided an unequivocal understanding that it is very much the conscious mind that drives our motivation. We are capable of using experience and personal qualities to direct our focus to reach our true potential. **You are in control of your own destiny!**

The interest in motivational research was prompted by the obsession of organisations that required their workers to be more productive in the thriving industries of the mid 20th-century, it wasn't until psychologist Dr Edwin Locke developed his now–famous goal setting theory (1968) that we gained a much deeper understanding of how to influence our own motivation.[5]

Although there are multiple behavioural theories surrounding performance and motivation (and we will explore many of these throughout the book), it is arguably the goal setting theory that is the backbone of why you are reading this book today.

Goal setting theory essentially tells us that specific and well–defined goals, designed with enough difficulty to challenge us, are the key features behind our motivation. Furthermore, goals serve an essential function, helping us to narrow our focus towards tasks that we see as relevant to our larger goals, and divert our attention away from distractions that would otherwise delay or stop us from achieving them. Having goals allows us to focus our mind and body towards them and perform at a higher level than we ever thought possible. This is something we innately want as human beings; to have the best lives that we can. Our dreams have the ability to drive us forward, energising our lives and allowing us to put in more effort towards our daily tasks. The short–term discomfort is often worth it to achieve what we want. However, our goals do much more for us than simply motivate us to work harder. Giving ourselves a chance to obtain our goals, our dreams, and the life that we want sets us apart as human beings. We want and need things for ourselves and our loved ones, and this need can be a powerful driver for our success.

The psychology behind the goal setting theory tells us something rather surprising: a challenging goal is more beneficial to us than an easy goal. Why is that? Well, because a difficult goal can bring forth extra effort and intensity, since deep down, we understand that we need that additional effort to achieve our goal.

Maslow's hierarchy of needs

In 1943, an American psychologist called Abraham Maslow described his original concept of the hierarchy of needs.[6] It is a motivational theory comprising seven layers (originally five, but was subsequently expanded by Maslow) of human needs, often shown as a pyramid that must be fulfilled for a person to reach their potential.

The general idea behind it was that, for a person to reach a level of personal and professional attainment, they must satisfy basic and more advanced needs, each building on the other, gradually creating a platform that motivates us to reach the highest levels. Initially, it was thought that the lower levels of needs within the pyramid, such as safety and loving needs, must be fulfilled in order to progress and attain further needs higher up in the pyramid, such as cognitive needs.

However, we now understand it to be a lot more flexible than that where we can occupy multiple levels simultaneously.

Maslow's hierarchy of needs

Initially, it was believed that each layer needed to be satisfied so a person could move on to the next layer. We now understand that a person can find themselves simultaneously in multiple layers throughout life.

What the levels mean

Physiological needs — At the very basic level, we need to eat, sleep, and drink, to survive.

Safety needs — It is natural to want control, health, and security, before we can really think about much else.

Love and belonging needs — As humans, we are deeply emotional beings, and the need to love and to be loved, for support, acceptance, and trust, are universal, and can give us strength in difficult times.

Esteem needs — This can be of higher or lower importance to an individual, but there is always an underlying feeling that we need respect for ourselves and from our peers, recognition of our achievements, and basic dignity in our lives before we can move onto other things within the pyramid.

Cognitive needs — On some level, we need to be mentally challenged and stimulated in order to improve and achieve a greater level of performance and productivity in our lives. Learning, understanding more about ourselves and our goals, and satisfying our curiosity about how and why things are the way they are, are all universal needs we each share.

Aesthetic needs — This level may vary depending on the person, but it describes the need for beauty and balance in our lives. To look and dress as we see fit, and which highlights our personality and quality.

Self–actualisation — The final level in the pyramid and the most difficult to attain according to Maslow's theory, and this level sees a person realise their potential, grow as a person, and for the purposes of this book, achieve their dream and life ambition.

Interestingly, some iterations also label the top of the pyramid with an extra level called 'transcendence' which would be the uttermost attainable virtue of human consciousness.

In addition to this hierarchy, Maslow listed certain actions and personal qualities that he believed would help someone to achieve this level. He explained how living a life of trying new things, being honest, working hard, and taking responsibility for your own actions would give a person some of the principal ingredients needed to reach their potential. These qualities are widely accepted as strong personal characteristics and ones that we have all witnessed in others and almost certainly possess ourselves.

While the hierarchy of motivation and unique qualities is a good starting place, we don't need to follow these to the letter. In fact, there is merit in creating your own hierarchy, using your own needs and desires to improve motivation and your chances of success. What's more, you can already identify the strongest components of yourself and others around you, and begin building a picture of the kind of person you want to be, and just as important, what type of people you want to surround yourself with. Use this as a way to think about these character traits for your own hierarchy as you understand what you value the most in yourself and what would bring the most benefit towards your goals.

Does the hierarchy work?

There has been a lot of scientific debate surrounding the precise layers of the pyramid, especially around whether each layer must be fulfilled in order to reach an upper layer in the pyramid. Current psychologists, and even Maslow himself, who later adapted and improved his hierarchy, now believe the pyramid to be much more flexible than initially thought. In reality, we often move up or down, throughout the layers in our lives, and are capable of occupying and fulfilling multiple layers in the pyramid at any given time. The overall concept behind Maslow's hierarchy of needs can be viewed in another way: we are all complex and unique people who rely on multiple needs and desires in order to grow both personally and professionally. Not everyone will move through this theoretical pyramid in a uniform manner, and that is OK, because at the very least, it allows us to look at the categories of needs that we all have so we can strike a balance between them and understand what we need the most and what we should prioritise.

More recent research has sought to test Maslow's theory in a cohort of over 60,000 people from 123 countries.[7] The investigators spent years conducting studies and analysing data and found that their results supported the basis for Maslow's hierarchy of motivation. They concluded that there are universal needs for humans that contribute to our motivation and overall performance in our lives regardless of age and cultural differences, and that this hierarchy encapsulates many of them. Although they agreed with the concept of the theory, they ultimately confirmed what many had suggested previously — that the levels should be defined by the individual and each person should rely on their personal hierarchy of motivation.

Seeing Maslow's hierarchy as a general understanding of what we need in order to reach our peak performance allows for times in our lives when things do not go to plan, when we may fall through these theoretical levels as work and life obligations fluctuate. This will happen on your journey towards your dream, and it is important to remember that it is perfectly normal to fall and take a step back, as long as you know that you can climb back up and climb even higher than before. Maslow's hierarchy has taught us that, if nothing else.

Maslow estimated that only 2% of people would ever reach the final level of the pyramid. This is your time and your pyramid. Reach the top of your own pyramid this year. Be a part of that 2%.

2

CREATING A BASELINE FOR YOURSELF

Every day is a fresh start. A new chance to become anything that you want to be.

So, who is this method best suited for? The short answer is: absolutely everyone! Whatever area of your life you want to improve in, this method will work with you to reach your goals and realise your dream. Your goals could be anything from learning a new language, reading more books, or starting a new business. It can also work for improving at something very active, like sports or fitness goals. Whatever your dream, *The Quantum Method* will work to break down your goals into smaller pieces that can be built upon during any part of your day. Over the next weeks, months, or years, you will see improvements. Some of these improvements may be difficult to visualise at first, but it is important to keep the bigger goal in mind and remain consistent with your hard work. This is why you will need to create a baseline for yourself; a series of questions and answers that help you to understand where you are in your journey and where you are going.

First, you must learn how to be as efficient as possible in this new system so you can maximise productivity with the bare minimum of change and sacrifice to your life. We do this by analysing your daily habits and schedule as they are today. It is for this very reason that *The Quantum Method* is so named, because it analyses the smallest components of your schedule to identify areas for improvement. These components are often overlooked in traditional methods, but they are the key to creating small but consistent gains over time. This method can work exceptionally well, but only when coupled with another component we will examine — time management. It is equally crucial to identify inefficiencies in our schedule early, because these are usually the best pockets of time to work on your goals.

How often have you organised your day with endless lists, only to be left feeling exhausted with an overbooked schedule? Or worse, you face the reality of not being able to complete everything in your schedule, so you cut time from somewhere else to make up for it.

The coupling of the analysis of your goals and schedule with time management is a powerful combination, and we will explore the scientific reasoning behind this and how we can use these strategies to work for us. But, as important as time management is, it should be noted that not all time throughout your day will be equally productive for you. Of course, there are always going to be times when you will just have to push forward despite feeling like you don't want to work on your goals, but we will work to minimise these instances.

All the talking is now done and we are about to begin this process;

now, we need to understand your preferred working styles. Below, you will be asked to take a look at your working habits and answer the following questions as honestly as possible. Even if there are some answers that you don't like, please remember, there is little point in cheating here; after all, it is only you who will see the answers. Please think carefully about each question because it will determine how to integrate the methods into your own life. It is never easy to look at yourself introspectively and critique your best and worst habits affecting your productivity. But remember, it is necessary in order to understand how to get the most out of the following chapters.

"THE GOALS YOU HAVE ARE FOR YOURSELF AND NOBODY ELSE, SO YOU SHOULD DREAM AS MUCH AS YOU WANT"

Write down the answers to the questions in your journal. If you are unsure of how to answer any of the questions, then feel free to take some time to write them. One way to do this would be to make notes throughout the day so that, by the end of the week, you will have a better idea of some of your answers. Don't forget, you can continually update your answers after implementing some of the methods for yourself, tailoring them as you learn more throughout the process. Don't worry if you are unsure about answering any of the questions: try your best to answer, continue reading through the book, and come back to them when you can. Reading through the remainder of the book will help you to have a better sense of the philosophy and approach we are taking. However, you will need to revisit the chapters once you have your answers to the following questions. Time to begin.

Question 1. When do you feel the most awake and alert during the day?

This is one of the most straightforward questions to answer as it will likely come to mind immediately. Rather than think of this time as simply the morning or evening, we are going to split up the daily hours during which you are awake, into two-hour blocks, ranking them from 1–8 (this could be more than 1–8 depending on how long you are typically awake for in a given day). Put a number 1 in the 2-hour block where you feel the most alert, and write the number 8 in the 2-hour block where all you think about is sleeping, and there would be little point in trying to get any work done here. The numbers between (2–7) should be written where appropriate. The bigger the number, the more likely it is that you want to stop and rest.

What is helpful about using the 2-hour blocks is that they can be placed in any order that you feel is right for you and can be viewed independently from each other, meaning they don't have to follow a numerical order. For example, just because you feel awake and alert between 9–11 am and fill it with number 1, doesn't mean that you won't feel tired and sluggish from 11–1 pm, and put a number 6 there.

For the best results, create numbered time blocks for each day of the week, rather than a general day. A recent report[8] suggested that our most productive times during the week are between Tuesday and Thursday, from 10:30 am to 3:00 pm. These values will vary daily because multiple factors contribute to our energy levels, such as our mealtime, exercise, workload, daylight, and stimulants like caffeine. To this end, if there is a particular time window when you have caffeine, exercise, or eat a meal, it may be worth jotting that down in the time window too. It will help to give you an idea as to the reasons you feel the way you do. For example, after carbohydrate–heavy meals, it is normal to feel tired and sluggish for an hour or two, and it is also normal to feel more alert after consuming caffeine (although not always, depending on your daily consumption and tolerance).

QUESTION 2. WHEN ARE YOU THE MOST PRODUCTIVE?

Although this question may appear very similar to question 1, the answers may not necessarily be the same. A scientific study looking at the productivity of people throughout the week reported fluctuations in cognitive function and productivity levels depending on the day.[9] In other words, there are many factors that can have an effect on our productivity and so we need to look at them individually.

For instance, you could feel awake and alert immediately after getting out of bed and taking a shower but have responsibilities that prevent you from working on your daily tasks and being productive. Perhaps there are children that need to get ready for school, a dog to walk, or a long commute into work. These are times when you cannot be productive (or so you think). It could also work in the opposite manner. For example, perhaps you finish work for the day and carve out an hour in the evening but are exhausted and ready for your bed. You can be productive during these hours because you compel yourself to work, despite feeling tired and needing to rest.

You can make another schedule for this question, again using 2–hour time blocks numbered 1–8, with number 1 placed in times where you are the most productive, and the number 8 in times when you are the least productive. Make sure to fill out all of the 2–hour time blocks with the remaining numbers. We will use both schedules from the first two questions to find the best time for you to be your most productive, but for now, separate the plans in your journal. In later chapters, we will look for overlap between the two numbering scales.

Question 3. How do you learn best?

From our experiences as children, most people assume that we learn because we watch someone stand at the front of a classroom and talk at us for hour upon tedious hour. For some, this can be a useful way to learn, especially for those of us who can listen attentively and leave with a newfound knowledge of a topic. For others, it can feel like white noise and a mind–numbing example of how not to learn. There is a wide variety of methods for people to learn new information and skills, and recent research is paving the way to explore these methods for individualised learning.

For this question, ask yourself about your preferred method of learning something new. For example, are you a visual learner? Meaning that images, videos, and diagrams integrate themselves into your memory better than a person explaining something or when you read from a book. Maybe you learn very quickly when you are actively doing something. This style may be suited for things like dancing or sports. Do you require visualisation and coaching before putting everything into practice to hit the ground running? This may require a slower and more methodical learning style, so you are absolutely clear on the task and expected outcomes before starting. There are many ways that you can learn something new, and taking a little time to understand what works best for you will pay off in the long–term, because whatever your goals are for your future, there will always be new and unfamiliar paths that you take. Like anything new, there will be a learning curve, and understanding how to get the best out of your time when learning them will improve your productivity in the long term.

Learning Styles

Not only do we all have our own preferred style of learning, but it can change depending on what we are learning. Feel free to play around with different methods to find what suits you best.

Let's shed a little more light on the different learning styles that you may want to have a think about. A verbal learner will respond well when ideas and concepts are explained to them, like in a classroom lecture or podcast, but also by reviewing written texts. The use of language will convey ideas more effectively compared to other methods.

Physical learning is one that occurs by actually doing something. A person who favours this type of learning might learn faster by performing the action themselves and figuring it out by trial and error.

Logical learning is typically associated with mathematical problems that need to be understood. The control and order of numbers and routines will appeal to the conventional logical learner.

Auditory learning is done by listening to information so it can be processed more clearly. This can range from audiobooks or podcasts to the typical musical notes of a song that a musician can understand and replicate often without the need for any written material.

Visual learners respond well to images, graphs, charts, and general illustrations that grab their attention to explain concepts. Mind maps are a common method that visual learners use, where all of the ideas and concepts are sketched out in front of them, making it easier to integrate the information and connected themes.

Social learning occurs best when information and ideas are communicated clearly between groups of people. This a common activity within classrooms or workshops where people are placed into groups who need to work together to solve a problem. Real estate, sales, and project management are professions that often attract social learners because teamwork and communication are natural methods for a person to learn more about their role and improve their abilities.

Solitary learners are more adept at taking their own time to work through any material that needs to be understood, because it allows them to find explanations themselves with few distractions. This is often done by reading through a manual or textbook, or simply working in a quiet space.

Despite the learning styles mentioned, it is more likely that you will learn by utilising all the different styles depending on the circumstances and information that you want to learn. However, some styles will appeal to you much more than others, and understanding these categories can help facilitate the learning process for you in the future.

Question 4. What are your short, medium, and long-term goals?

This will be covered in more detail as we continue, but it is a good idea to have them in your mind throughout every chapter. After all, they are your reasons for reading this book and learning how to become more productive in the pursuit of your goals. I'm sure everyone has heard of these types of goals before, and so it is not about reinventing the wheel here. For this question, you just need to write down some of the main goals you have. Later in the book, we will explore the science behind goal setting and how to maximise your productivity by using a more nuanced approach to goal setting than you have seen before.

To start with, let's take a look at the short-term goals. A short-term goal could look something like, "I need to finish reading this book by Friday for the book club I am going to this weekend", or "I need to run 20 miles by the next weekend". Something that needs to happen within the next few days or weeks would be categorised as a short-term goal.

A medium-term goal could be anywhere from a few weeks from now, to six months, and could look like, "I want to learn how to have a basic conversation in German", or it could be, "I want to build a website and blog that will complement my social media account within six months". Like with all of the questions here, try not to write the first thing that comes to mind. Dig through your thoughts and put effort into describing your goals. You can have as many or as few as you like, so don't be afraid to be ambitious.

The long-term goals are something entirely different. Typically, they are anything from 6–12 months or longer. They are usually what we think of when we talk about our goals and dreams for the future. It could be something like, "I want to be financially stable by the time I start a family in the next five years", or "I want to be world-famous for my art 20 years from now". It could literally be anything you can imagine, no matter how strange or far away that idea may seem today. Remember, the goals you have are for yourself and nobody else, so you should dream as much as you want without the fear of how others may perceive it.

With that in mind, although you are under no obligation to tell anybody your goals and dreams, there are benefits to letting others participate in them. This can be done to hold yourself more accountable or to simply share the joy you find in working towards your passion. A large research study that combined the data from over 140 smaller studies with a total of 16,500 people found that goal setting had a significant ability to change a person's behaviour and improve performance and productivity.[10] What's more, goal setting became more effective when the goals were difficult, facilitated group work, or were shared publicly. Therefore, the benefits of sharing your goals with others around you and updating them on your progress has been demonstrated to be significantly beneficial to your goals.

Question 5. What motivates you best?

I encourage you to have a long think about this one. As always, remember that there are no wrong answers here. Knowing where our motivation really comes from can take some personal digging to find out. We might never have thought about this question before and found motivation naturally and without any effort. Motivation can come from anywhere and doesn't necessarily need to be in one place or make any statement about your character. For example, you could be motivated by money, but not by greed, because you want a higher quality of life. Or you could be motivated by the attention and praise you receive for your hard work because it makes you feel good about yourself and validates your efforts. Perhaps you are motivated by short–term rewards and gifts like new clothes, video games, or vacations. Another common motivation is a person's family, and how the extra work they do can eventually lead to improved circumstances for them. Ultimately, if something encourages you to work harder and put in the extra effort required, then it is a perfect motivator for you. It is the end result we want, so feel free to write down anything that comes to mind. This would be a good moment to think back to Maslow's hierarchy of needs and how you began to create your own idea of what you need. Feel free to update and refine your hierarchy, and indeed, any answers to these questions, as you read further.

Motivation is great at the start of your journey and once you learn more about your own motivation, it becomes easier to develop. Still, it is unlikely you will have a continuous stream throughout your progress. You should also think about how you will be motivated six months or five years down the line. One way to think about this part is to ask the question: If you achieve milestones along the way, how will you celebrate? Are you the type of person to celebrate the small victories and progress you make, or are you a person who prefers to keep moving forwards toward their next goal? Motivation may be different to the feeling of being successful or appearing successful to your peers. Instead, it might be that you are motivated by metrics of improvement, like the amount of weight lost in a diet and exercise goal, or the amount of money earned each quarter, for a specific sales goal. For this person, it would be better to take a more analytical approach to your goals and achievements, which could be one aspect of the motivation in itself.

Think about how you can motivate yourself throughout all stages of your journey towards achieving your goals and keep updating them in your journal as you discover more.

QUESTION 6. HOW WILL YOU MONITOR YOUR PROGRESS?

The answer to this question, unsurprisingly, will be different for everyone and unique to yourself. It depends entirely on the goals and dreams that you are going to achieve.

For this question, it is recommended that you set monthly self-feedback sessions. Times where you review what you have achieved in the previous month and how far you have progressed. We will explore the scientific reasoning behind this feedback later in the book, but as this is an introductory questionnaire, it is best to not overload it with too much detail just yet. We want to build on the concepts as we read more and the best way to do that is to start with a more straightforward outline and add detail as we progress. The same is true for any answer you write down for these questions. Starting with a basic idea that you improve on later is a great way to begin if you are not sure how to answer them completely just yet.

It could be that your particular goals require a more extended period than a month to see progress, but for at least the first three months, you should hold these review sessions. They will be incredibly valuable for keeping your focus high and allowing you to identify any challenges and obstacles that may repeat themselves in the coming months. These sessions also help to visualise your path forward. It can be easy to look at the bigger picture and feel discouraged by the size of the task in front of you. When we stop to look at how far we have come and how we have progressed towards achieving our dream, it can be a powerful method to keep motivation elevated and remind ourselves that any time you work on achieving that dream, you are taking one step closer towards achieving them.

Over time, you can change the frequency of these review sessions and customise them to suit your schedule, but for now, try to use them at least once per month. In addition, think about every possible way you could monitor your progress and write them in your journal. Many of these chapters require you to test out what works best for you, and an excellent place to start is with multiple ideas, slowly honing in on what works for you. Of course, you can use any system that you feel comfortable with that will provide you with the most benefit. However you decide to monitor your progress, use this part of the book to write a list of around 15 ways to monitor your progress each month, and as the year goes by, reduce those 15 to a smaller number that is more helpful and meaningful to you.

Question 7. How will you know when you have achieved your goals, at precisely what point?

This question is the most fun to answer because you get to imagine yourself living your dream and visualise it in your mind. The precise moment when you achieve that dream could be anything. For example, if you want to buy a house, then it could be the act of signing the contract that is your final accomplishment and signal that you have reached your goal. Perhaps it will be turning the key in the front door and seeing it open for the first time. If you want to publish a book, finding it on the shelves in a bookstore will be worth visualising. If you want to improve in a sport, then perhaps signing up for a competition and placing in the top 10 would be the time, or better yet, winning it. You can dream as big as you want!

Whatever the event, you should try to be very specific. The exact moment that you will know. This is because you can visualise that moment in your head. Not only the event itself, but how you feel, who will be there next to you, or how you will share the news with your friends and family. Make it as real as you can. Picture everything and keep those moments safe in your mind. This picture will probably change as your journey evolves, but you should always have a precise moment in your head somewhere.

QUESTION 8. WHAT IS YOUR DREAM FOR YOUR FUTURE?

Lastly, it is time to write your dream and goals down in your journal or notebook. Write as much or as little detail as you like, even one sentence will work. Write it in big bold letters to signal your intention. Sign a mental contract with yourself that no matter how long it takes, you will achieve it. **You deserve it!**

One more thing

There is one last point to note before we continue with the following chapters. *The Quantum Method* is a great tool to help you become more productive in your day, but it is just that, a tool. Without the person wielding it, this method alone will only take you so far unless used with consistency, dedication, and focus. Keep in mind, that in order to improve your productivity and work output, there are always some sacrifices to be made and a new level of focus. I will always encourage you to dream as big as possible, but we also need to be grounded by the level of hard work and consistency required. But that is why you are here. You already have the desire; we just need to focus it to maximise your effort. Let's do this!

3
EASY, MEDIUM, OR HARD?

I can do a thousand things but only one at a time. My value lies in understanding which one.

In order to reach your goals, there are a number of things you need to do along the way. Primarily, there are tasks that will need to be completed, which, when finished, will combine together to achieve your goal. The important message to take away from this chapter is this: **not all tasks are created equal.**

In this chapter, we are going to discuss how you can group your tasks into three categories: easy, medium, and hard. These categories will be important in the chapters ahead, because they apply to any task you have and can help to structure your day on a different level than you are used to.

The way we start is to create a list of all the tasks that you might need to complete, and then break them down into categories depending on the time and effort needed for each one. Once you know that, it will allow you to determine how best to designate your spare time to each of them. Don't worry if you think you don't have much free time to offer; we will also look at how to extract more from your day throughout the book.

The first step is to use your answer from question 8 to state what your dream is that you will achieve and then to make a list of smaller actions and steps that will lead to it. It is natural for this to seem daunting at first, and it is very likely that at this stage, you are probably unaware of exactly everything you need to do. Don't worry; the list doesn't need to be exhaustive today, because you can add to it as you go through the book. But for now, try to write at least ten tasks that would have to be completed over the course of this month in order for you to feel productive. A to-do list of things that you know need to get done. For instance, if you want to become more productive in your current job, your tasks could be anything from making calls, sending emails, conducting meetings and presentations, to data entry and generating reports. Write down whatever comes to mind and keep the list close so you can revisit it whenever new ideas come to mind.

The categories

When we describe grouping your tasks, think of them as three very long to-do lists. They might only have ten items today, but eventually, they could have hundreds of items filling them. Some can be done quickly without needing a great deal of attention, and some will naturally take longer and require your undivided attention. By understanding which ones fit into the categories below, we can group them into selected times in the week when they can be completed with the minimum intrusion into your regular daily schedule. It is no use trying to run 10 miles in preparation for a marathon when you only have 15 minutes at lunchtime to do it, right? The task wouldn't fit the schedule. But there are tasks that can be done in those 15 minutes which would improve your marathon time; strengthening and stretching muscle groups could do that, for example, as it would work towards the larger goal, even in 15 minutes.

Easy

To begin with, what exactly are easy tasks? These are any items on your to–do list that can be done at almost any moment you are free as they generally don't require high concentration levels. We often do these types of tasks throughout the day and refer to being on 'autopilot' for many of them. General examples might include making repetitive calls, data entry, writing up notes, or updating a word document, but there are many more.

Tasks can also be grouped into the easy category based on time. Writing an email, for example, might require you to think carefully about what you need to say and double–check spelling and grammar, but if it can be done in less than 30 minutes, then it will fall into the easy category. This means you can place the task of writing an email into a time window throughout your day when you have a short opening. Sending the email from your phone while waiting in line for a coffee would be an example of this. Tasks that can be categorised as easy tasks are more susceptible to distractions, particularly at home. Therefore, if at least some of the easy–category tasks can be done remotely (like using your phone when waiting in line) when you are multitasking, then even better. Tasks can be completed before the opportunity for distraction presents itself.

One more note on emails. Another study looking at productivity found that after sending an email, people took 16 minutes to refocus on the task they were previously doing.[11] Although that time might be different for you, it is a statistic that is worth keeping in mind when creating your schedule. If you know the following 16 minutes are not going to be productive then fill that time with another easy task. Let it become less about how you feel in the moment, and more about the automatic behaviour of starting on the new task.

Medium

As a good rule, the medium difficulty tasks should be things you already know how to do, with little requirement to think creatively or learn something new, although this may not always be the case. These tasks will require more attention and effort, usually between 30 minutes and 2 hours. This category depends on your goals, but you will learn to recognise how to place your tasks into this group over time, and we will explore some examples of medium tasks in the following pages.

Hard

Finally, we have the hard tasks; these demand your maximum effort and undivided attention. They are the tasks that may require original thought, patience, or skill, and generally take the most time out of your day. Tasks that are placed into the hard category are usually the ones that come to mind when you think about achieving your goals. Tackling hard tasks can make you feel like you are making significant progress and are typically noticed by others around you. Want to sell more houses? Selling a home would be a hard task. Want to grow your business? Meeting with a potential client and agreeing on a new contract would be a hard task. Want to get into shape? Being in the gym would also fall into this category.

You will likely find that most tasks fit into the easy category, fewer in the medium group, and even fewer in the hard category. This might come as a surprise, as we often place more emphasis on the hard category tasks because it feels like we are making the most progress; but in reality, nearly all day–to–day tasks can be broken down into smaller, simple activities.

The organisation

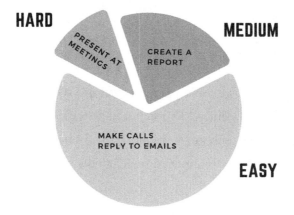

The majority of our daily tasks can be grouped into the easy category once we break them down into smaller chunks. Surprisingly few are grouped into the hard category.

Don't worry if you misplace tasks into different categories, as they can be moved around whenever you need to. If you are unsure of where to place a task, put it in a more difficult category and downgrade it later if you need to. Surprisingly, task complexity does not negatively affect your ability to reach your goals.[10] This is important, because it allows us to create as many smaller goals and tasks as we want, and to set our schedule up to include more time-consuming and difficult tasks without being concerned with our ability to complete everything and reach our goals. Formulating different strategic goals and task lists is, therefore, worth taking the time to work on.

Another way to think about this grouping system is to imagine it as a traffic light, with green (easy), amber (medium), and red (hard). The green tasks are easy, and you can drive right through them without too much effort. Amber tasks may need you to assess the situation but are often easily accomplished if you devote a little more time to them than the green ones. Red tasks require you to stop, think about your next move, and assess the situation in front of you.

This might be good in theory, but what does it mean when put into practice? Let us use an example of how you can break down a project from start to finish. Although this specific example may not be identical to your chosen goals, try to think about how your individual tasks will fit into the categories in the example below.

A story of success

In this section, we are going to put what we have just learned about grouping tasks into a little more context. What better way than to use a real-life example of how the quantum method was used by a friend of mine back in 2017 when I was working towards my PhD in neuroscience, and only a year after I had finalised this method?

One February lunchtime in London, UK, I found myself discussing my plans for the coming year with my friend, Thomas, who told me of his desire to learn a new language. Eventually, he wanted to move to Spain with his Spanish fiancé Luisa and build a life together. The only problem was, he didn't speak the language. He knew only a few of the basic words and phrases, courtesy of Luisa's patience, which wasn't going to cut it, and he was certain that he would need a firm grasp of the language to continue his career in Spain.

"This is perfect", I told him. A smile spread across my face, much to his bemusement, as I explained that he could put my new method to the test, and essentially, be my guinea pig. I knew how powerful the quantum method could be, mostly because I had been employing some version of it in my own life for many years. It was helping me through a PhD in neuroscience, teaching myself to play the guitar, and climbing to the top of high-altitude mountains all over the world. But this was the first time I had explained my method to someone else, and definitely the first time I was brave enough to bring it into the world for others to see.

After a couple of weeks of discussion, we sat down together and came up with a list of things that he would need to do in order to learn and practice Spanish. We also went through the list with Luisa, so she could be included in his journey and help in any way that she could. We gave ourselves the extraordinarily ambitious goal of learning the language in only one year. After which, he would take a Spanish language exam, and apply for jobs in Spain with his new qualification. Adding the time frame to the goal enabled him to really focus his attention to it, because that deadline would be getting closer every day.

The great thing about learning a new language is that there are a lot of resources already out there for beginners. This made it much simpler to compile the list of tasks and group them into categories, because there was always some new way to learn Spanish. Once we had this list of tasks, we built on them in more detail. We will go through this in each chapter of the book, but to give you an overview, I will explain a few of them now.

Often, we made sure to use flashcards and frequent quizzes,

because neuroscience research explains that this retrieval process can stimulate neuroplasticity — the scientific term for learning and memory — in a much more significant way than other methods.[12] It was also crucial to space out the learning rather than cramming it into one long day or weekend. In some cases, zonal working like this can yield excellent results, particularly when used for up to four hours at a time, but for more academic goals, frequent studying in smaller doses is much better.[13]

We also looked at every part of his schedule and decided where his task categories could go. His easy category tasks were saved for the evenings or lunch breaks. Through more than a little trial and error, Thomas found that his medium category tasks would be used randomly throughout his week as he saw fit. This worked well for watching movies in Spanish or writing notes for Luisa around the house. They made a game of it. The harder tasks were generally placed at weekends, but because he lived with Luisa, he would always have opportunities to practice.

The next step was to find suitable gadgets for him to learn on. Laptops, phones, books, flashcards, and online videos were all used, depending on the task and his personal preferences. That way, wherever he was, if he suddenly felt the urge to practice his Spanish, he could. He particularly enjoyed listening to Spanish conversations on his phone when exercising on weekends. He even had Luisa record some phrases for him so he could practice while exercising.

After a couple of months, we checked in with each other to see how his learning was going. By this time, he had tailored his tasks to suit his daily schedule, and after some initial trouble finding the energy to practice in the evenings, his schedule was optimised. There was one issue, however: by the time two months had rolled around, the initial excitement had decreased and he was finding some parts of the routine monotonous and tiring.

This is where the idea behind Chapter 8: Goals and Rewards, was developed. Thomas' problem was that all he seemed to be doing was working on his Spanish. Day in and day out. That is a great way to get results in the short-term, but some of the initial fun was being lost. This was when I introduced him to Level 1, 2, and 3 goals and rewards. We used short, medium, and long-term goals that he could check off frequently. Crucially, each goal came with its own reward. Whether that was drinks with friends, or takeout food for dinner, or sometimes a more significant reward like a weekend away, it could inspire him to keep working hard. The harder the task, the better the reward. After all, there is a real need for some downtime and rest, even when being as productive and consistent as possible. And guess what? It worked. Thomas found that enjoying frequent small rewards offered him a chance, not just to bask in his Spanish-speaking glory, but to take a look back at his progress and feel great about the reward and even better about seeing how far he had come. He later told me that he marked on his calendar every Level 1 goal and reward (for him, this was at the end of each month) and would get more excited the closer the target day came.

To cut a long story short, at the 12-month mark, and as promised, Thomas took a Spanish language test in order to receive an official language grade that would help him apply for work in Spain. He passed easily with a grade that would enable him to apply for his new job. His hard work paid off! Around six months later, he made good on his promise to move to Spain with Luisa, and it was very much a 'happy ever after' story. When writing this section of the book, I asked permission to share his story, and he excitedly told me that he had something else he wanted to share. He is now happily married and spends many wonderful nights with his family and other locals from his community, all speaking Spanish with him.

An example of the categories we used

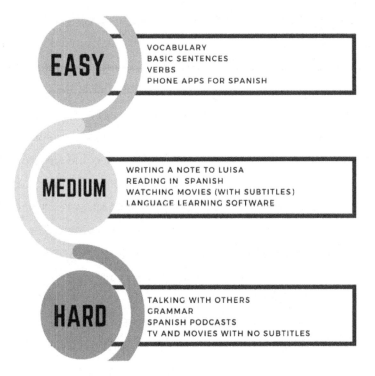

It may take a little time to optimise the placement of each task into your chosen categories. Ultimately, there are no wrong answers, as it is whatever works best for you.

Plan, plan, and plan some more

Hopefully, you have seen how all tasks, large and small, hard and easy, all contribute to the bigger piece of the same puzzle, even if some achievements will not be appreciated until months or years later. Any dream or goal you have for your future is always far more complex than it appears at first. You can try and plan as much as possible, but there is always a multitude of things that arise throughout any journey that can interrupt your plan or affect your rhythm. This is another reason why these categories can be so powerful. If your schedule unpredictably changes, you can swap your planned activity with another from the same or perhaps an easier category. This way, when something unexpected happens in your day, you literally have a list of quick and simple tasks you can do. In moments like this, it is best to pick another task that is easier than you had planned. A harder category task will be more difficult to slot into your schedule on short notice.

Now that we have a clearer understanding of how to group your categories, it is time to revisit your list of tasks from earlier. Go through each one and label them with either easy, medium, or hard. We will use this list throughout the book and build on the detail for each category, explaining how we can find time throughout the day for each of them.

Below is a quick example of how tasks from the earlier example would be categorised:

What are the categories?

	START/STOP	PREVIOUS EXPERIENCE	AUTOPILOT	TIME REQUIRED	TIME TO SHOW BENEFIT
Easy	YES	YES	YES	<1 HOUR	3-6 MONTHS
Medium	YES	OFTEN	OFTEN	<2 HOURS	3 MONTHS
Hard	NO	NO	NO	1-4 HOURS	1 MONTH

Use this table as a guide to create your own categories. Remember, they can be updated as often as you like, so they do not need to be perfect right now.

The Overview

The next steps

1. In your journal, write out the tasks you might have to do in the next year.

2. It doesn't need to be very specific; for now, only a general list of approximately ten tasks.

3. If you can group them into categories, great; this will give you an idea of where you will place them within your schedule.

4
UNDERSTANDING YOUR TIME AND EFFICIENCY

Our time here is not infinite. That is the true beauty, because we must make each moment count.

Time management is nothing new. Modern time management as we know it became popular in the 1970s with a series of books that would go on to inspire countless more over the years. However, we need to be a little cautious here, because time management alone will not give you everything you are working towards, but it will be advantageous when used appropriately.

In addition to breaking down the components of your goals into more manageable pieces, *The Quantum Method* is also about using your time as effectively as possible. On a basic level, this is a time management system, but as we build up your schedule throughout the chapters you will see that it is much more. The first part of this chapter will explain some of the hurdles you will face when attempting to improve your productivity, things like bad habits and procrastination, and explain the psychological reason for why we do. The second part of this chapter revolves around improving your own schedule and identifying how inefficient your current schedule is, so that it can be restructured and improved.

In order to build a time management system, it is essential to understand what exactly comprises time management and how it can be evolved and integrated into your schedule to work for you. From data collected in research studies, we know there are three main components to a good time management strategy, all of which are utilised within this book.[14]

1. Structure: We will focus heavily on structuring your tasks around your schedule in a way that significantly boosts your productivity without dramatically adjusting your home or work schedule.

2. Protection: If you have ever had a plan to work on your goals but ended up doing something else, then your original time was not protected from distractions. Here we deal specifically with the biggest distraction of them all, procrastination. We will discuss its causes along with strategies you can use to remove procrastination from your life.

3. Adapting: Flexibility in your plans is inbuilt into this book to allow you to adapt to a demanding and changing schedule while remaining productive. Your plan will be flexible enough so you can change or swap specific tasks at short notice, and you will learn to give yourself feedback on how it works for you. It is also important to update your schedule throughout your journey so that you can build on the things you enjoy and get the best results from.

"YOU CAN THINK OF TIME AS AN INVESTMENT FOR YOUR FUTURE LIFE"

The overall consensus from over 150 scientific studies is that a well thought-out time management strategy can not only increase productivity, but also boost self-esteem and all-round performance.[14] However, the true benefit of utilising time management comes from understanding how it can work *for* you.

At their core, most time management plans target a process called 'activity maximisation'. It essentially creates a schedule where the greatest number of tasks are completed in a given time period. This book also incorporates this strategy to find immediate improvements to your productivity. This can have immediate benefits because research shows that when we plan our schedule with a focus on activity maximisation, the probability of completing our tasks significantly increases; specifically, when our schedule includes *when* and *where* we will complete the tasks.[15]

What many plans fail to appreciate is that there is a more important benefit of time management, which is 'outcome maximisation'. This shifts the focus onto making each task work towards your ultimate goal in some way. Rather than completing lots of items on the to-do lists, each item is only included if it serves a specific and identifiable purpose for the bigger goal. Moving forwards, the following chapters within this book incorporate both the activity and outcome maximisation methods. This is a more profound concept than it might initially appear, because when we tell ourselves what our dream is, it can feel like a huge and overwhelmingly distant ambition. By incorporating outcome maximisation into this book, we can break down the goal into smaller sections and target them individually at different time points.

Opportunities wasted

It is incredible how much time we all waste on our devices every day. I admit, I have also been guilty of this. We all do it. Our phones, social media, YouTube, music, computer games, Reddit, TV and Netflix, all of it! That is fine if we don't want to work to our potential, but if we're going to increase our productivity and bring our dreams into a reality, we need to take an uncomfortable look at our bad habits. Think about it for just a moment and I am sure you will be shocked at how much time is spent on things not directed towards reaching your goals.

Before we go any further, I want to stress here that these outlets are perfectly fine to use. Of course, they can also be valuable tools to help with your business or social life, but too often, we get distracted by them. It is important to state here that although we will be exploring how much opportunity actually exists in your day, it is also perfectly acceptable to have moments where you need to relax, do something completely opposite to what you might see as productive, and generally de-stress. There is no problem with recognising those instances when you genuinely need to take a break. Taking time to re-centre yourself and remember why you are doing this, are valuable tools, and well worth the investment. The issue is more about the quantity of time we spend doing nothing, and how honest we are with ourselves about our motivation to do the work.

Time scarcity

50% of us experience time–scarcity; not enough time in the day![18]

15% of adults struggle with procrastination![19]

This more than doubles amongst university students![20]

Time scarcity is likely something that we have all noticed but not become aware of in such a detailed way. Hopefully by identifying it in our daily lives, we can learn to manage it better.

Procrastination

Procrastination is a time thief! It is one of the reasons why we are not where we want to be today and why each of us struggles with achieving our dreams. Of course, there are other factors, but procrastination plays a pivotal role. Essentially, procrastination is the time spent distracting ourselves with other things rather than doing what we should do to achieve the future we want.

It is important to understand your own level of procrastination in order to improve productivity and your quality of life. Procrastination has been linked to stress, depression, and excessive anxiety,[16,17] contributing to lower productivity. We need to address procrastination if we want to improve productivity to new levels.

Why do we procrastinate if we know it is harmful to our progress?

If we want to overcome procrastination, an excellent place to start is by appreciating why our brain procrastinates in the first place. It has been suggested that procrastination had an evolutionary benefit for us as human beings, as it may have evolved as a by–product of another character trait — impulsivity.[21] Impulsive behaviour probably benefitted us throughout human history when we needed to adapt quickly to changing environments without the need for long–term planning. However, in modern society, acting impulsively can have its drawbacks. When we act impulsively, we do things that are not aligned with our goals, and when we do this often, it is procrastination. Essentially, multiple impulsive decisions made over time culminate in procrastination. To improve our productivity, we need to think about how our actions will contribute to our ultimate success, and recognising when procrastination and impulsivity are detrimental. When we fail to overcome our impulsive behaviours and act on them, we find ourselves doing things in the moment rather than focusing on our current task. Procrastination may have originally been a way to focus on short–term goals by ignoring long–term goals. Even though they may have had evolutionary benefits, when we live with the desire to improve our performance and productivity to attain long–term goals, procrastination is our enemy.

Current research on the causes and effects of procrastination has identified several reasons for it, such as fear of failure, low self–confidence, inadequate planning, and perfectionism.[22] However, planning alone may not be enough to overcome procrastination, because the act of planning itself, when used excessively, can often be used as a tactic to tell ourselves we are working when we are not. The key is to acknowledge that procrastination is an issue, one that is fairly normal, but not allow it to creep into our schedule.

Overcoming procrastination

As a means to overcome procrastination, we need to shift our focus to the process of improving productivity, that is, identifying areas within your schedule where you can improve productivity and how to utilise devices and technology to improve further. When we focus on *how* to pursue our goals, we use scientifically proven methods to reduce the fear of failure by changing our focus from the bigger objective, and therefore, unknown potential consequences, to a much more micro-level view of whatever task is ahead of us today.[23] In other words, when the feeling of wanting to procrastinate occurs, we have already laid out the blueprint of how we will work. It is now a case of following through with your pre-structured plan, rather than finding yourself in a situation where you have time and a general list of tasks but no real sense of how or why you will complete them.

The *why* is also crucial in defeating excessive procrastination. Temporal-motivation theory[24] was developed by two psychologists, Piers Steel and Cornelius König, to explain how time is a big motivator. The theory suggests that procrastination may also correlate with a person's goal setting ability. This means when our *why* is not clearly defined, we tend to procrastinate more. This makes sense because if we know what needs to be done but don't have the value of those tasks (the *why*), then distractions and procrastination can feel as (or more) valuable than the tasks themselves. If we can vividly picture the end result of our dream, and understand that our actions will lead us there, it can reinforce the reasons why we need to stay focused and procrastinate less. This effect becomes even more powerful if we can attach an emotional context to it. For example, will improving your productivity, and therefore reducing procrastination, improve your financial stability in the future and help support your family or loved ones? Why do you have your goals, and why is it essential to strip away any residual procrastination?

Taking both of these factors together, time management interventions have been shown to have significant benefits in reducing procrastination, and this is why the methods within this book can be helpful to anyone who struggles with procrastination.[25]

Self–confidence has also been identified as a trigger for procrastination. Having confidence in yourself and your ability to achieve your dreams is not an easy feat. It can take a long time for that belief to set in, and doubt can be a lifelong struggle, even with success. If, at any point in your journey, you find yourself unsure whether you are capable of reaching your goals, then do not worry. Decades of studies into human behaviour tell us that people draw from their personal experience and unique transferable skills that have helped them in similar situations. In other words, you already have the skills you need to accomplish your dream, even if you do not realise it yet. Remember, you have every right to feel confident in your abilities. If you didn't have what it takes, you wouldn't be reading this book. You would not have decided to take a chance on yourself and start your journey towards your dream. Having more confidence in yourself (even if you have to fake it for a while) will have a direct influence on your level of achievement by raising your standards and limiting procrastination.[26]

How to start

Here is what I would like to ask you to do. Feel free to do this as much or as little as you want, but I would suggest investing some time and effort into this assessment as it will feed into many of the other chapters in the book, especially when we come to finding the right times when you will work on your different daily tasks.

The first step is to break up your day into three segments. This will depend on your own schedule for precise timings, but it will look something like this:

1. The morning from waking up–12 pm

2. Afternoon from 12–5 pm

3. Evening from 5 pm–sleep

Think of them like three miniature days in one, because we can structure each of them independently. A great feature of doing this is that on days where you don't quite manage to get everything done, say, in the morning, you still have two more attempts (afternoon and evening). It doesn't feel like the day is lost because it didn't go to plan in the morning. You get three attempts to have a productive day. William Durant, the 20th–century creator of General Motors, said it best when he exclaimed that we should "forget past mistakes, forget failures. Forget everything except what you are going to do now and do it!" So, if there are moments in your day where you feel like you didn't work on something you had planned, then it is OK. I encourage you to think of this quotation and move past it towards the next great thing you will do.

Time inefficiency

Firstly, what is time inefficiency? Let's look at it this way: are you very productive until lunchtime, and then the afternoon lag comes and you find yourself scrolling through your phone more and more? Are there multiple breaks of five minutes in the morning and then an evening of time–wasting? This will be important to understand, but as mentioned earlier, it is not about removing all of this time, it is about finding a healthier balance that fits into your own life and goals. Here, we only need to have a general overview of how much time is being lost. So, total up the time in each segment of the day, to give you an understanding of the total time that is wasted. Make notes on your phone or journal with estimates of how much time you could have been working on your tasks.

Examining your day in this way is a useful exercise at this stage because it produces a clear–cut example of how much time really exists in your day. This does not mean that we will look to remove any shred of time you spend relaxing, of course not, because relaxing and allowing yourself time to take the pressure off is also important, but at this early stage in your journey, we need to see just how much time is there in the first place.

Building on what you know

For the next seven days (make sure to include weekends), note down any time you spend doing something that wastes your time, or to put it a better way, uses your time inefficiently. You could write this down in a journal or anything you have with you on a daily basis. There are mobile phone applications that work really well for this. If you spend 5 minutes on YouTube, or a few minutes checking your social media feed, then write it down. Perhaps it could be 45 minutes listening to the radio on your daily commute or sitting in a car for an hour waiting for your child to finish their after–school activities. Whatever it is, make sure you write it in your journal. If you are unsure if it needs to be written down, note it anyway, and you can always change it at a later date.

"Give yourself credit for doing this. It is courageous and challenging".

This is done for a week rather than a single day, because each day will be different depending on your schedule. For example, Mondays may be particularly busy for you, whereas Wednesday and Thursday evenings may be when you have free time and have an opportunity to work on your goals. Once you have your week of timings written out, double-check the time inefficiency items on the list to make sure you are comfortable that everything is included.

Time to cut

When you feel ready, you can use what you have learned. Start the next day by cutting this time by 50%. This means that you can still have moments of distractions, but they are limited to around half the time you would normally have. Over the following months, we can reduce this even further, although I would limit this to a maximum of 80% reduction, because you don't want to burn yourself out, and as mentioned earlier, having time to relax during the day is recommended. So, if during the day you spend one hour on irrelevant things that use up your productive time, look to reduce that by around 30 minutes. When you catch yourself on your phone, social media, or whatever it is that distracts you, acknowledge it and start a new task from the list of 'easy tasks' you have created.

This may feel like a big change, but it has been shown that during our time spent at work, only 60% is spent doing things that would be deemed productive (probably best not to mention this to your manager).[27] If we apply this statistic to our home life and free time, we can easily see how there is room for improvement. If you want to be more productive in your profession and are motivated by career advancement, then keep this statistic in your mind as you aim to cut time inefficiencies in your day.

These timings will change as you move closer towards your goals and become more familiar with the process. You will need to assess them often at first to find the right balance for you. So don't worry too much if it feels unnatural right now; it will be a dramatic change to your schedule that takes time to adjust to, but it will become easier over time. As you learn to integrate *The Quantum Method* into your life, you will begin to identify areas of your day that could be segmented differently, and you will be free to make those adjustments to your own schedule.

It is worth noting that you may not get the balance perfectly right during the first week of working with 50% less time inefficiency. It will require constant monitoring and feedback from yourself. This can be achieved more easily now we have broken the day into thirds because you can use each segment separately, and any time you catch yourself wasting time, stop yourself and have a task at hand for you to work on, preferably an easy–category task. Over the course of the next few weeks, you will find it easier to reduce your time inefficiencies. You will be amazed by how much you can improve your productivity with this one technique alone.

How to use it

The two steps discussed are enough to help you to become aware of the extra time throughout the day that you probably didn't realise you had. We will build upon this idea later in the book as we discuss different devices that you can use to work on your tasks in different situations you encounter. Now that you are aware of the time that is wasted, eventually, you will need to fill it with the easy, medium, and hard tasks you have already listed. For now, it is enough to only tag those times with the tasks you could do. So, if you have ten minutes at lunch time when you typically use social media, you can find an appropriate task to perform instead. Once you have steps 1 and 2 written out, try to get creative with what you could be doing instead. This way, when you catch yourself wasting that time, you already know what you could be doing instead, and can begin working on that task.

Use the diagram of time inefficiency to help you. The more critical areas are steps 1 and 2, but they can be analysed further if you want to dive deeper, as shown in the diagram on the next page.

Example of time efficiency

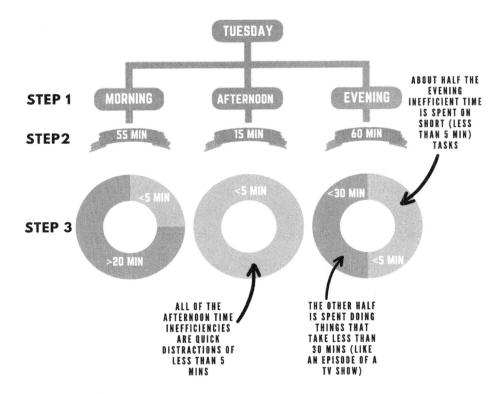

It is incredible how much time each day we lose. Once we identify this time we can begin to mark out segments that we can use to become more productive.

Inefficiency schedule

For those of you who want to go a little deeper and extract even more time from your day, there is a way to do it. Once you have your inefficiency schedule, as I like to call it, you will need to break down each segment into estimates of time, which is shown in step 3 of the diagram. This means that each time you do something, like using social media, you should try to have an idea of how long you will spend doing it. For example, are most of your inefficiencies around five minutes? Or are they longer, maybe 30 minutes or more? Are they varied or dependent on the day of the week? You do not need to be exact here, but it should give you an idea of the types of free time you will have in each part of your day, once we clean up some of these inefficiencies. We will use those time slots to give an even clearer idea as to which productive task can be substituted for these moments. A useful function of smartphones is the ability to look at daily screen time. Your phone will even break down the total screen time into specific apps, making it easier to gather this information. A simple look into the phone settings may yield valuable data for you.

If you are someone who finds themselves struggling to use your new time effectively, then don't worry. It can take time to learn to 'switch on' to boost productivity when there are only short windows of time throughout the day. It has been reported[28] that it can take people up to one hour to 'ramp up' to take on a productive task. The very thought of it can be daunting and we often need time to adjust. This is fine to begin with, but as we go through this book, we will learn to switch easily into productive mode whenever we need to and reduce procrastination to zero. Let other people take one hour, and while they are wasting time, you are getting your tasks done, taking one step closer to your dream. Keep this in mind the next time you have a short time window you have worked hard to identify in your schedule, and make sure to fill it with productive tasks that are going to benefit your long-term goals.

Remember to give yourself credit for doing this. It will feel like a drastic change to your daily habits, but it is incredibly inspiring when someone actively changes and improves their lives. It is courageous and challenging, and it already says a lot about your strength of character that you are willing to do it.

The Overview

We all spend a lot of time throughout the week doing nothing. At least, nothing productive that is going to help us to reach our goals. By identifying how often we do this, we can find moments in the week to work within, thereby increasing productivity without changing your normal schedule.

The next steps

1. Draw out your own diagram using step 1 to break down your day into three parts, and step 2 to identify the total time wasted in each part. You can use the template on the next page as a guide.
2. Then, begin to fill each of those times with more productive tasks in an effort to reduce your time inefficiency by at least 50%.

A template for your own schedule

When we learn how much time we have throughout the day, it can make it easier to decide which tasks could be placed there.

5
LEARN WHAT WORKS BEST FOR YOU

I know myself better than anyone else ever can, and that is a great strength of mine.

The Yerkes–Dodson Law

An important principle that fits nicely with the concepts behind this book — the Yerkes–Dodson law. Proposed in 1908 by two psychologists, Robert Yerkes and John Dodson, it describes the relationship between stress and task performance. Essentially, we work at our optimal level when we are under just the right amount of stress or time pressure, and we perform best when we find the tasks exciting and engaging. Too little stress in our schedule and we are prone to boredom and distraction. Too much stress, and we can feel overwhelmed, rushed, and anxious about how we can manage everything.

Think about this for a moment. How often have you been in your job and spent days working with little productivity? Had a weekend when you did nothing but binge on TV? I'm sure we all have! Now think of moments when you had a deadline approaching or a project that had to be finished. Naturally, you would have been more productive in these moments, not only because it was required, but because as humans we reach our pinnacle when there is a certain level of pressure and mental stimulation. Often, in these times of productivity, we can work so hard that we are unaware of the time passing; this is sometimes referred to as being in 'the zone'.

Think of other times when the work, whether professionally or in your home life, has built up to the point where you cannot picture how everything will get done. Standards drop, tasks get forgotten or abandoned, and we are left feeling exhausted, broken, and emotionally drained.

How we make the Yerkes–Dodson law work for us is to find the optimum level of stress for you personally. Some people may find they can tolerate large amounts of pressure, and even thrive in those situations, while others may find themselves more prone to anxiety and feeling overwhelmed or inadequate. All of these situations are perfectly acceptable, because the idea here is to find what works best for you and nobody else. It may take some trial and error, but the Yerkes–Dodson law is an interesting concept that can be used, if for nothing else, to understand how you work at your peak level. This can be used in terms of setting deadlines for yourself or announcing your timeline to friends or family. It could also be incorporated into your journey by creating a workload that is high enough to leave less room for unproductive days.

Whether or not the initial principles of the Yerkes–Dodson Law, which were founded on giving mild electric shocks to mice, represent an accurate picture of what is happening inside the mind, it is undoubtedly an intriguing foundation to build upon. As part of your process within this book, it will be up to you to decide what level of stress and pressure you want to create for yourself. Whether that is time pressure, such as deadlines, a financial strain, such as earning a specified amount of money by the end of the year, or simply, the pressure of reaching the target you have set yourself. Too much pressure, as is apparent from the Yerkes–Dodson law, will lead to performance deficits and, even worse, anxiety and mental health issues that are in no way worth enduring in the pursuit of your happiness and fulfilment. Too little pressure and you will be letting yourself down because you know that you could achieve more if you push yourself further. Over time, learn what works best for you and what you find to be the most fun and engaging, and make a note of it in your journal. Eventually, your schedule will be tailored to exactly what you need to perform at your optimal level. Just remember to adjust it frequently based on your performance feedback.

Looking forward

The extra time you have carved out of your week from your inefficiency schedule will now be used as a framework for this next section. Here, we get into the real meat of the technique, where you will use some of the answers from the questionnaire in the introduction to understand which times during the day are best suited for each of your tasks. This will be based on your alertness scores from question 1, which give you a general idea of when you will be working at your best.

We also need to take a look at your answers to question 2, where you have a breakdown of your week filled with 2–hour blocks rated from 1–8, with each number corresponding to how productive you are. If you have not done this, it might be worth pausing for a moment and thinking about this type of schedule.

"Make sure to personalise this method around your own unique life."

How to use the scores

We can now take the information you created from your 2-hour time blocks and put them to use. In the previous chapter, you broke down your day into three segments of morning, afternoon, and evening. Use the 2-hour blocks from the questionnaire to understand which parts of your day are more productive for you and when you are most alert. You can use the diagram below as a guide.

Do the 2-hour blocks for high alertness all fall in the evening? Are your 2-hour blocks for productivity generally in the morning? Perhaps they are mixed and vary throughout the day.

We use this chapter and your scores to give you more context about the three segments of your day, thereby building up the level of detail on your own version of the daily time efficiency schematic. The 2-hour blocks can be used to determine when you could work on an easy or hard task, but they are generally more useful for identifying trends in your day and week. As bestselling author Stephen Covey once wrote, "the key is not to prioritise what's on your schedule, but to schedule your priorities". In other words, we need to find a way to make the schedule work *for* you and identify the areas within it that are suitable for your tasks.

Step One

Add in your numbers explaining how alert and awake you feel. Remember that they do not have to follow consecutively and can be placed in any order you like.

Task categories

Next, look for any time window you rated 5–8 (remember that the larger the number, the lower the alertness). These times are suited for moments in the day when you are not particularly alert and awake. Easy tasks should be assigned to numbers between 6–8 for alertness. For the scores of 3–5 in your alertness, it is best to place medium tasks here. Of course, it is also acceptable to have some deviation in the two numbers here, especially if you feel strongly about grouping them differently. As always, make sure to personalise this method around your own unique life. The hard tasks should only be reserved for times in your day when you are both alert and feel like you could be very productive. Tasks in this category need to work around your schedule a little more than the others because they are more intensive. Scores of 1 or 2 (your most alert) will show you where to place hard tasks. Naturally, this is time–dependent on your work and home schedules, but it can serve as a useful system.

Use this scoring system as a guide to learn more about how you prefer to work, rather than having it set in stone. It has the flexibility for you to inject a little creativity into your own schedule if needed. Use the diagram below as a reference for how the task categories can be placed into the schedule and make sure to do this for every day of the week.

Step Two

Now it is time to place the different task categories into each time window. Save the harder tasks for when you feel more alert.

Productivity and alertness

The best results will be found when your productivity ratings overlap with your alertness scores. If they do overlap, it means you have a time window each day where you can really make progress. Take a moment to overlay your productivity scores from question 2 with your alertness scores from question 1, so you can look for times when they are similar or, hopefully, identical. They may not necessarily overlap, as a person can be productive in a tired and fatigued state if that is the only time in the day they have to work. If this example is true for you, even though you are productive, I would recommend that you change your workload here to an easy or medium task regardless of the high productivity rating. This is because your best work will never be achieved when your body and mind are at their lowest. However, if you find an overlap, make a note to prioritise that time window on that specific day. So, for example, if on Tuesday night between 8–10 pm you have scored yourself as both highly productive and highly alert (with scores in the 1–3 range) then try to prioritise Tuesday nights for your goals and any hard category tasks you have. You can work throughout the week just like normal, but put a reminder in place to ensure that Tuesday nights are cleared for you to work on whatever tasks you need to get finished. It will show the greatest benefit for you. This is when you can use the scores as a more precise micro–level guide, rather than the general overview.

Step Three: Combine them all

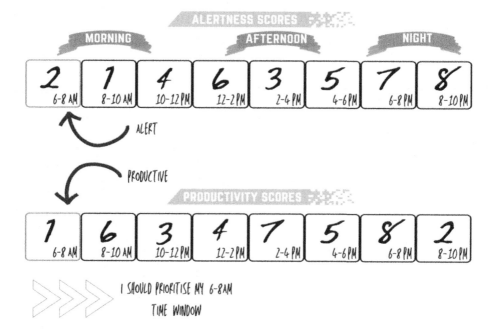

Here is what the time blocks look like when we add the productivity scores. Often, it may be the case that the two scores do not overlap because of other obligations, but if you can find a time window within the schedule, it can be extremely useful to identify at this early stage.

Life happens

Busy home and work lives will interfere with your time windows, which is why it is preferable to use this chapter as a general guide to understand your working habits better. There will be many instances where you cannot operate towards your goals during your most optimum window. That is a natural part of life and does not mean that you can't be more productive in other ways. Instead, look to the time window where you can be productive, perhaps with another task, and utilise that as best you can. The most productive and alert times may be reserved for weekends or an occasional day once or twice per month. That will just be a consequence of your regular schedule. The quantum method is designed to mould itself to your schedule and doesn't have to replace it. So, using what we have learned with the lists of tasks, alertness, and productivity scores, feel free to play around with different tasks during various time windows you have identified. When you look back and realise how you succeeded when other people didn't see the chance, that will be an incredible achievement. There is no magic wand; instead, we have a detailed look at how you can extract the most from your day. Remember that it is these small consistent efforts that reap the rewards. The idea here is to demonstrate how to assign the workloads to each part of your day.

Another point to mention is the amount of time needed to work within the time blocks themselves. Do you need to work on your tasks for the entire two hours? Not really. With our work and life schedules, that isn't really going to happen. In an ideal situation, the two hours would be used solely for your tasks and this may occur from time to time, particularly on weekends when you may have a little more time. In reality, however, you will need to find short windows of time within these 2-hour blocks in order to make this work, and this is what we will focus on in the next chapter.

Earlier in the chapter, we mentioned the Yerkes–Dodson law — that a more engaging and intense workload can boost performance. The Yerkes–Dodson law is particularly applicable when deciding where to place your easy, medium, and hard tasks. That is because the more routine and straightforward the task, the more it can be scheduled for times of high stress, meaning that if you have a certain period of time where you have pressure from your schedule in your work and home life, the easy tasks will be an ideal group to be placed here. We are capable of completing easy tasks even when we feel the pressure from other parts of our lives, so that we can be as productive as possible, regardless of your schedule.

Feedback

We have regular check-in points within *The Quantum Method* because a vast volume of research teaches us the value of feedback in our accomplishments.[29] If we do not know how we are performing, we cannot make the requirements needed to excel. We need to understand what is working, what isn't working, and how to improve. At the end of each month, spend half an hour making notes and thinking about how your schedule is working for you. To this end, it is essential that we ask the following questions in your monthly feedback sessions:

- What tasks bring enjoyment and which ones don't?
- Which tasks are being completed and which are not?
- Which tasks were easy to fit into your schedule and which were difficult?
- Which tasks need to be placed in another part of your schedule?
- What task would you like to do more?
- What could be done to improve in the future?

This feedback will help to finely tune your schedule depending on your results. Remember, this system is personalised to you and your style of working, and so it requires feedback from yourself in order to maximise your progress. External feedback from milestones and peers, or other people close to you, will have an additional benefit. In general, this type of feedback cannot be done at one time and forgotten about. It requires committed feedback throughout your journey because your skills and experience will change as you develop, and adapting to those changes will provide consistent boosts to performance.

Using feedback for motivation

You may find that on some tasks, you are underachieving, and in others, you are exceeding your targets. Use the feedback questions in the list above to adjust not only your effort level but also your goals. For example, if you find that you are consistently missing your productivity goals, then use that as further motivation to ensure you hit your targets next time or reduce the target to a more manageable one to ensure success. Think of feedback as a way to monitor goal setting because creating your goals is a continuous cycle of learning and evolving to keep your productivity high. Even though you will write down your goals at the start of your journey, never be afraid to alter them based on your feedback and results.

"FAILURE IS NOT DEFINITIVE, IT IS SIMPLY A METHOD BY WHICH WE LEARN TO ALTER OUR BEHAVIOUR FOR SUCCESS"

Scientific studies which looked at the influence of feedback on performance have explained why this type of extra motivation in the face of failure might occur.[30] As humans, we innately want to remove anything in our lives that we feel disappointed by or prevents us from reaching our goal. Therefore, using this feedback system to understand, not just the overall success or failure, but also the specific areas of success or failure, will further enhance motivation.

In social psychology, feedback methods like those in this book play into something called 'attribution theory'. Developed in the early 1990s, attribution theory describes how we interpret positive and negative feedback, or, to put it another way, success and failure.[31] Receiving feedback is not an easy thing to do. We can feel embarrassed or attacked when we hear something that goes against our view of ourselves and our achievements, even if that feedback comes from ourselves. If everything was positive, then that is great, and we can happily accept it and move on. However, to truly reach our potential and achieve our dream, we also need to feel comfortable with receiving negative feedback and analysing what didn't work for us. Furthermore, attribution theory explains that we tend to attribute positive feedback to our own abilities and negative feedback to other factors that are out of our control. This effect is so powerful that we can even change our opinions about other people depending on the feedback we hear about ourselves.[32] It puts us in a good or bad mood when we need to look at something objectively.

For this reason, when you hold your feedback sessions be sure to do them at a time when there are fewer distractions, and you feel relaxed and optimistic. As always, try to think about all of the suggestions and changes you might need to make, then take a few hours or even another day before you finalise any changes you will make, especially when based on negative feedback.

The Overview

Feel free to use this template to build your own version and identify which segment of the day should be dedicated to each category of task.

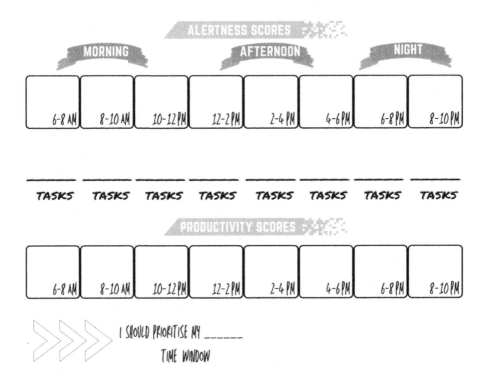

With a layout like this, we can add all of the information together and identify the times throughout each day where we can be productive. We can prioritise time blocks where we can find an overlap between alertness and productivity. In the absence of any overlap, at least we can now visualise the schedule and begin to place our tasks into each time window.

The next steps

1. Create your own schematic using your alertness scores and divide the day into three segments. Do this for all days of the week.

2. Decide which category of tasks could be placed into each time block throughout your day.

3. Find potential overlap with productivity and alertness and prioritise that time for your most demanding tasks.

6
ZONAL WORKING

Given enough time and patience, we have no limits.

As you may have already noticed, the quantum method relies on planning and analysis of your schedule in order to extract small but consistent benefits from undiscovered time. While this is helpful, there are instances when planning only takes you so far and it is time to focus on a single activity.

In this chapter, we are going to introduce a method of working that will show immediate benefits and progress towards your long-term goals. It can be used as a standalone technique for instances when you have an abundance of spare time, or it can be incorporated into the alertness and productivity scores you developed in Chapter 5 and used for shorter segments you have identified as highly productive.

The concept is relatively simple. Zonal working is when you give yourself a number of tasks to accomplish in a very limited time frame.

Let's say that you have three separate tasks that must be completed by the end of the month. On a Saturday morning, you would take three hours and divide them up into 55–minute time windows. Each window would be given the sole focus of one of those tasks. For example, if you are trying to grow your business, the first 55 minutes could be spent emailing potential clients and following up with existing contacts. The following 55 minutes could be spent writing a new pitch for further investments, and the final time window could be used for research on another market for new infiltration.

> # "BACK–TO–BACK SCHEDULING CAN MAKE THE REMAINING TIME FEEL LONGER AND ALLOW YOU TO BEGIN ACTIVITIES THAT MIGHT OTHERWISE FEEL MORE SUBSTANTIAL"

Although it can be used in conjunction with Chapter 5, zonal working is best used on a day when you have at least 3–4 hours of uninterrupted time, for instance, on a weekend morning. It may not be ideal for everyone, as it can be challenging to carve out this time in an already busy life, but even if you can only manage it once per month, you will notice its impact very quickly. Zonal working is best used in a 4–hour window with up to 90–minute intervals; this is based on research by Professor Anders Ericsson at the Florida State University, USA, which has shown that quality and productivity can drop when we engage difficult tasks for longer durations.[33] This would suggest that zonal working with medium–hard category tasks should be kept to a maximum of 4 hours per day, although easy–category tasks could be fitted in throughout the remainder of the day if you feel so inclined.

Surprisingly, the power of zonal working does not come solely from the time you dedicate, but rather from the fact that it forces you to give your undivided attention to each task, with the knowledge that your time is limited. The clock is always counting down from the moment you start working in your time window. Yes, you can go for a walk, make yourself a drink, or check your phone, but that timer is constantly counting down and you know that once it hits zero, you will need to stop and move on to the next task. After a little practice, it is incredible how much you can get done in that time when you know there will be no additional minutes added on.

To start zonal working, all you need is a timer. If you are using your phone, it may be worthwhile to look for an app that allows a Zen state that essentially blocks the phone from use (except for emergency calls) for a set period of time. You can set it for 55 minutes and rest assured that you will not be tempted to use the phone during that time. This method of blocking phone use is particularly effective, because research investigating nearly 7,500 people who had difficulty with procrastination confirmed that proximity to the distraction (rather than just having the distraction anywhere) had a considerable influence.[34]

Something to keep in mind is that the time windows in zonal working doesn't have to be 55 minutes. It can be any amount of time you want, as long as you can keep focused and don't drift off into distraction. Further studies by Professor Ericsson also stated that cycles of 90 minutes of high concentration could yield optimal performance. Adding a short break between those time zones can also help.

Taking breaks

Most of what is contained within this book will focus on getting the maximum out of your day so you can work as efficiently as possible without dramatically changing your schedule, especially through zonal working. Despite this, it is important to offer a gentle reminder that breaks are allowed, whether that break is an hour, a day, or even a week. If you feel like you genuinely need it, or you are struggling in some way, then taking a break to recharge and reset is absolutely fine. Taking regular breaks gives your brain a chance to relax and reset for further tasks and improved consistency. There is data to suggest that regular breaks improve both quality of your work and productivity,[35] so factor in breaks within your schedule. The key here is to ensure that you are taking the break for the right reasons and not using it as an excuse to ignore your commitments. Your enjoyment when working towards your dream, your mental health, and your ability to consistently perform at a high level, will all benefit from taking a break when you feel you really need one, rather than when you simply want a break.

As a cherry on top, taking a short break has been scientifically proven to be beneficial to your day. Psychological research looking at potential methods to boost concentration[36] and creativity[37] have shown that taking a 50–minute walk can lead to impressive benefits, particularly when the walk is engaging and mentally stimulating. Unfortunately, this means that running an errand or walking around familiar roads by your house is likely to have only a limited effect, but walks with interesting features or surrounding nature will work well. Even viewing a picture of nature can influence a person's creativity.[36]

Although it may not be possible to find these environments close to you, if the option presents itself, give yourself permission to incorporate them into your life and recharge your mind in order to come back stronger.

You decide how to structure zonal working

Ultimately, you decide how you work most effectively and how you want to use zonal working to help you. Perhaps you prefer to get multiple tasks done throughout the day, potentially four blocks of 30 minutes. Or maybe you would prefer to focus on one or two tasks for three hours each, particularly if you need to be active or travel for them. At first, try zonal working with the timings suggested above using 55 minutes, and over time feel free to adjust them to suit you.

It is also worth keeping in mind that each task can be weighted differently. If one task requires two hours and the other only 30 minutes, you can designate the correct amount of time for each. It is more important that you change to your new task at the end of the time zone to keep the momentum going, rather than how much time you allot for each of them. The fast pace of zonal working will help to keep you interested, motivated, and productive because the knowledge that you have limited time will keep your focus high.

When you have finished the 55–minute window, take a 10–minute break and set up your next task. After the break has passed, turn on the timer and start again. Even if you feel you could keep working on the first task, it is time to change to task two. This is for two reasons: First, finishing a task but still regarding it with excitement for the chance to continue it will keep your enthusiasm levels high for the next time you do it. If you associate the act of doing the task with a positive feeling, over time you will associate that task with positive feelings, which helps you to stay motivated for longer.

The second benefit of moving on to another task is that once the day is over, you can look back and see your progress in multiple areas, which has the added psychological effect of ticking off more tasks from your to-do list and seeing your progress towards your long-term dream, and incorporates the activity maximisation discussed earlier.

A small caveat

A quick note here. I just mentioned that it is preferred, at least for the first few times you try it, to move on to the other tasks once the time has been completed. I do believe that will work for the general reader. That said, and although I swear by zonal working for my own goals, if you truly are in a great rhythm and cannot fathom the idea of stopping, then allow yourself some freedom to continue. It does go against the general concept of zonal working and should be used sparingly, but I also believe that every structure needs to have inbuilt flexibility. Try to stick to the allotted time, but if you must keep going on a specific task, you have the flexibility to make that decision.

THE OVERVIEW

Use your scores from Chapter 5 to find moments to use zonal working. This technique uses blocks of time with no distractions to create a focused, goal-driven time to work solely on your tasks. It may not be easy to find the time in your schedule but try your best to use this technique at least once per month. It will be worth it.

The next steps

1. Look over your scores and schedule from the previous chapters to determine when you have a block of time between 1–4 hours.

2. Divide it into smaller segments and have a clear idea of what you will work on during each time zone.

3. Find a place where you are not likely to be interrupted or find other distractions.

7
ANYTIME, ANYWHERE

Every day is an opportunity. I either take it or I don't, but every day I don't, someone else might.

The neuroscience of decision making

When we make decisions, the brain uses a combination of experience and a risk : reward ratio, to help us predict future outcomes of those decisions. This involves neural circuitry in regions of the brain that work to reinforce certain behaviours. Sometimes that could mean repeating old behaviours such as habits or trying new ones.

Eventually, these brain signals make it to regions in the frontal cortex, where our logic and prediction areas are found, but in the process, multiple regions throughout the brain become involved including the cingulate cortex, locus coeruleus, prefrontal cortex, orbitofrontal cortex, hippocampus, and many others. While it is not necessary to go into detail for each region involved, it is important to appreciate that there is a complicated network within your brain guiding you to make the right decisions in your life about everything, from what you will eat for breakfast to how you will work towards your goals on any given day. They are all subject to this decision–making progress.

It is important to have a basic conceptual understanding of this process because we can appreciate how complicated decision–making can be. This complexity can lend itself to a huge number of influences that can affect our decisions, particularly ones that involve our goals. Therefore, we should simplify the decision–making process as much as possible so it is not susceptible to one of the hundreds of opportunities for bad choices to creep in. The trick is to place the decision–making process early in your journey, so that later on, when you feel tired, unmotivated, or otherwise distracted from your tasks, all of your decisions and scheduling have already been made. Our brain will want to distract us from difficult and energy–sapping tasks based on how we feel in that moment, and so we must create our workload at a time when we feel more objective.

There are two important pieces of information that, above all else, we should be aware of:

1. We do not always make decisions based on our long–term success. We make decisions primarily according to what will give us an immediate reward. This was beneficial during our evolution when vital resources were scarce, and we needed to prioritise our short–term survival over anything else. But today, we must engage more of our experience and motivation so we can structure our plans based on what will be best for us in the future, such as achieving our goals. If we create a

schedule based solely on how we feel in the moment, we will not be as successful as when we have a pre–formulated program that we created at a previous, more emotionally neutral time.

2. We can gain a better understanding of why experience plays a role in our decision–making and why we should strive to push ourselves in new environments. We can exploit this mechanism of experience–based decision–making by writing down our feedback and using it to help us guide future decisions.

Anytime, anywhere

So far, we have discussed creating extra time within an already busy life. We have broken down the day into segments and identified times in each of them that are being used inefficiently. Here, we will build on these two themes to explore how to use the time windows you have identified wherever you are. As we now understand how our decision–making can benefit from planning out tasks appropriately, this chapter can help you to plan your schedule in even greater detail.

Unlike zonal working, which requires longer blocks of time, here we want to uncover how you can become productive in short periods throughout the day. Have a look at your list of tasks and categories you created earlier in the book. We can use that list to determine how to complete each task using a number of different devices and platforms.

The basis of this chapter revolves around the idea that, to increase your output, you will need to find methods by which you can work on your tasks wherever you are and in every type of situation. We do this by asking the question: What platforms do we need to use to accomplish each task we have? For example, can we use things like mobile phones, desk telephones, videos, computers, online meetings, laptops, or books?

In a way, it is similar to the concept of multitasking. Technically, the human brain can't multitask the way we think, like reading and writing at the same time. This is because each activity uses the same parts of the brain, such as the language areas. However, the brain is very good at multitasking when the tasks use different parts of the brain, which is why you can listen to music while jogging or have a conversation with a passenger while driving a car. Motor tasks (things that require movement) and cognitive tasks (thinking tasks) can be coupled together. We can use this concept to discover how you can find ways to improve your productivity at any time during the day or night.

Let us use one example here: say you want to grow your new business by sourcing products from another country to sell and distribute in your own country. There are hundreds of different moving parts to this type of business, from sourcing your suppliers to marketing and sales, distribution, accounts and tax, and many others that we don't need to go into today. Using this example, we can discuss what you could do to remain productive anywhere and at any time. First, we would refer to the list of tasks and categories previously created. Perhaps you would need to source products, email clients, research pricing, logistics, update a website and anything else that you can think of. Now that we have a brief list, the next step is to break each component down further. For example, could emails be done on a phone or tablet instead of using an office and computer? Could you download or print out product research to read wherever you go? In other words, how many different ways are there to do each task, and are there multiple devices that could be used to complete each one?

Take your time with this list and feel free to complete it over several days, adding to it whenever something new comes to mind.

Eventually, it will look something like the table below. Although this list is basic, it conveys the concept that there are many different devices and methods to achieve a single task, meaning that you don't need to be stuck in an office or on location to work on your goals.

Anytime, anywhere

Which platforms can be used?

	COMPUTER	PHONE	PODCAST	OFFICE	LAPTOP
Online Research	✓	✓		✓	✓
Client Calls		✓		✓	
Reading	✓	✓			✓
Client Meeting		✓		✓	
Learning A Skill			✓		

Can your work be done using multiple platforms? Explore as many different methods of working as you can in order to find the optimal tools for you to improve your productivity. Be as creative as you can because you don't know until you try.

The idea behind this is to identify any way that you can become more fluid and flexible in completing your tasks. If you can use your phone for something rather than a computer, you could prioritise your time in the office for tasks that can only be done when you are there.

Combined with your efficiency schedule from Chapter 4, you can find the moments in your day that could be assisted by the use of specific devices. For instance, this could mean that your phone or tablet would be useful while walking to get lunch or waiting for someone while sitting in your car. They can be carried easily and used for easy tasks that you have already identified. Go through your list of tasks and group each of them with potential devices you could use to help you be more productive.

Put it into action

Now that you know what you will need to accomplish your daily or weekly tasks, it is time to think about how to fit these tasks into your day. Again, this is where the time efficiency schedule can help, along with a good understanding of how your tasks can be grouped into easy, medium, and hard categories. We now also have an idea of the additional tools we can use to complete each task, so let us look at how we can combine everything into your new schedule.

Let's say you are waiting for a friend to show up at a restaurant. Are there tasks that you could be doing on the mobile phone while you wait? Phones have the capability to use PDF files, spreadsheets, word documents, video calls, and a long list of others, so let's use them as often as we can. From the example we are using, this would give you time to research items online, return emails, and briefly add to your research of competitors. You don't need to complete each task, only move your progress forward, especially if the phone can be synchronised with a computer.

Other examples could be:

- When you are waiting for a bus or taxi. Perhaps you could bring out the reading materials you carry with you to annotate them for later.

- Need to speak to clients over the phone? Could you use your cell phone and a headset and talk as you walk the dog, drive your car, or walk through the grocery store?

- If you are on more casual terms, perhaps speaking to a friend or business partner, could you talk on the phone while walking or indoor cycling?

Ok, so the last one could be tough, but the concept here is to change your mindset and unleash your creative side when trying to think of methods to squeeze tasks into parts of your day that you might never have considered could be productive. Each scenario will be specific to your own tasks you are trying to complete. Some of the more difficult tasks may need a little extra consideration, but the concept shouldn't change. Throughout the next week, try imagining the times that you could be doing something productive if only you had the resources to fit it in at that moment. Is there a podcast or audiobook that offers insight into your chosen industry? Perhaps you could listen to it during your commute to work. All of that time will add up to a substantial investment over the course of a year. Don't underestimate the volume of topics covered with these formats. This type of multi–tasking will be suited for the easy–category tasks, but there are plenty of opportunities for medium–type tasks too.

"It is these small increments that will lead to success"

The general focus is on how these small changes can be adapted to suit your personality, schedule and goals. If nothing else, this chapter should be used as a catalyst for the idea that there are multiple times in each day to work towards your goals that don't need to be viewed in a conventional manner. The possibilities are endless, and the best way to really utilise this method is to think creatively. Take the time to think about how you can incorporate different activities into your life to contribute to the bigger goal and dream you have. It is these small increments that will lead to success.

The Overview

Now you have identified times throughout your week to work on your goals, you should start thinking about how you can use them. Think creatively about how you can use technology (old and new), like your phone, books, tablets, podcasts and audiobooks, to aid your productivity.

The next steps

1. List the tasks you need to complete.

2. Break them down into the devices you could use.

3. Group both the tasks and gadgets and use it as a guide for making your schedule for the next week.

8
GOALS AND REWARDS

My limit is only temporary. Once I reach it, I will search for the next one.

The science of goal setting

Goals are important to us. They provide us with the opportunity to realise how our highs, lows, and everything in between all matter if we can use that experience to achieve the dream we have for ourselves.

Earlier, in Chapter 1, we discussed goal setting theory, which explains, amongst other things, how challenging and specific goals enable us to utilise the skills and knowledge that we have acquired over our lives and to help us reach new limits of performance and productivity. In this chapter, we are going to incorporate this theory into our own goals and discuss how to structure different types of goals to maximise their power. The structure of this book uses the underlying premise of goal setting theory to keep your motivation and performance at a high level regardless of where you are in your journey, because there are always new goals and milestones that you can achieve.

You might be asking yourself right now, "Sure, the theory is there but does it work, and will the goal setting in the quantum method bring me better results?" The short answer is, absolutely yes!

A recent study found that goal setting played a crucial role in promoting sustained behavioural change in an individual's performance.[38] Meaning that it wasn't just a quick alteration in their routine to accomplish something quickly, but goal setting allowed people to reap the benefits in the long–term.

That same study also found that the productivity and performance boost from goal setting could be further improved when each goal was tailored to the individual, accounting for a person's ambitions, their desired goal difficulty and level of ability. This has already been incorporated into the quantum method by regularly using feedback to update your goal setting and ultimately, to improve your productivity. Importantly, there is room for personalisation and restructuring of your schedule so you can tailor it to whatever works best for you.

Another large study looking at nearly 6,000 people using goal setting to improve both their commitment and performance in a task found significant benefits,[39] and that a combination of short–term and long–term goals was most effective for increasing performance and overall benefit. This was particularly effective for fitness goals, but it will work for any purpose you can think of. Research studies such as this are the foundation for the Level 1, 2, and 3 goals described in this chapter. We want to gain the most benefit from how we structure our goals and rewards, and use peer-reviewed scientific methods to make sure that happens.

So, how do we set goals for ourselves, and which type of goals do we need to get the maximum benefit? Firstly, the goal needs to be specific. Of course, we can have one big dream that we focus on, such as becoming financially stable, growing a successful business, or achieving an academic degree, but research tells us that this alone will not be enough. What does work is when we break down the larger dream into smaller and more specific goals that we achieve along the way. Essentially, we need to set short, medium, and long–term goals. Goal specificity reduces a lot of variability in performance by reducing ambiguity about what needs to be done to achieve your goals.[30]

In addition, spacing out these goals with specific deadlines evenly throughout the year will have an increased benefit on your productivity, because this allows a person to remain at a consistently high level of performance without feeling overwhelmed and anxious about too many approaching deadlines.[40] To this end, a study with nearly 3,000 people, looked at the difference in performance and goal achievement between two different groups.[41] One group was given goals which were not specific and were aimed at simply working harder to get a better result, such as 'do your best'. The second group had clearly defined goals whereby they needed to reach a specified number of completed tasks by a given deadline. The second group outperformed the non–specific group by a significant margin.

In short, specific and challenging goals consistently lead to much stronger performance in people who normally do not set any goals for themselves or set a goal of simply 'doing their best'. The do–your–best goal allows for too many variables that could derail the plan, and it is not capable of providing a single target to focus on. On the other hand, specific goals have been consistently proven to enable us to achieve more and ultimately bring our dreams into reality.[30]

Your goals

Refer back to your journal where you answered questions 4–6. We will use your answers as a guide for this chapter. As with everything in the quantum method, we first need to break down our goals into smaller and more manageable segments. The reason for this is that if we think only about the bigger dream, it can be discouraging when it takes a long time to achieve it. However long it takes, it doesn't mean you are not making progress. When you see Mount Everest 100 miles in the distance and walk 10 miles closer to it, the looming mountainous landscape will not have changed much in that time, even though you are closer than before.

Monitoring your progress is a simple and integral part of your process, because it is reassuring to look back over the previous time and recognise just how far you have come, and how much closer you now are to achieving your dream. The best way to do this is to have multiple types of goals, and then, of course, multiple types of rewards when you reach them.

Specificity and your goals

Level 1 goals

Highly specific Level 1 goals are ones that you will achieve in the short–term, generally between 2–4 weeks, and should be as specific as you can make them. For these goals, it is not about the bigger picture, only the short–term goal. This keeps motivation and performance levels high because it is always clear what needs to be done and how to obtain it.

Think of Level 1 goals as a traditional to-do list for your short-term goals. In fact, I encourage you to write down your short-term goals in a list, somewhere that is visible and definitely one that will be checked at least once per week. This will help to keep you focused on the weekly task to achieve these goals.

Don't forget, just because they are short-term goals doesn't mean that they should not be categorised by difficulty, or you should not break the tasks down further to understand where you can find time in your day to work on them. Try to incorporate all that you have read. The only difference here is that rather than working towards your final dream, you are using your new tools to work towards a smaller goal that is already in your sights.

Bonus tips for Level 1 goals

- Make sure to complete every task on the list and try not to move them into another week, even if sometimes they feel like a chore to complete.

- Share your progress with someone. A weekly update will let your friends and family know how well you are doing and provide you with a feeling of accomplishment.

- Make a note of completed tasks that you enjoyed the most. This is helpful because you can incorporate them into your schedule more often in the future. After all, working hard still needs to be fun (most of the time).

- When you complete your Level 1 goals, write one or two sentences in your journal explaining what you have accomplished in that time. Keep hold of them to refer back to and see your progression over the months. When you are feeling tired and hit a low point, you can use these descriptions to remotivate and encourage yourself.

"Take a moment to remember how hard you have worked, because you deserve your rewards"

Level 1 rewards — The fun part

Of course, completing your goals and making progress towards may be a reward on its own, but it shouldn't stop you from taking a moment to enjoy the achievement. Reward yourself after each Level 1 goal that you achieve. It doesn't need to be anything costly or extravagant, as you still have a lot of work to do. But try to take the time to treat yourself. Maybe that means a dinner out with friends and family, a movie night, or a simple relaxing bath one evening. Whatever you think is appropriate to feel positive about your progress and remind yourself of the hard work you have put in over the last few weeks.

The reason we have Level 1 rewards is because cognitive neuroscience tells us that the closer we are to completing a task or receiving a reward, the harder we work and are more productive we become.[42] By incorporating Level 1 goals and rewards, we take advantage of our natural strengths to focus and motivate ourselves to achieve what we need to, and early achievement can be an effective way to increase overall achievement.[43]

Level 2 goals

We will use Level 2 goals to remain specific enough to our tasks, while allowing for growth in the months leading up to them because each Level 1 goal will complement each other and contribute to the Level 2 goal. The spacing of Level 2 goals should be evenly distributed, around 3–6 months apart.

Level 2 goals were created for the quantum method based on research showing that when we create time landmarks, or 'fresh starts', we tend to have short but highly effective performance boosts, particularly when starting a more challenging activity.[44] Therefore, we incorporate Level 2 goals because once you achieve them, you have the opportunity to set a new goal, which will act as a catalyst to improve productivity as you raise your limits to face the new challenge.

Small bites of the pie are usually enough to finish the meal, but we can't always rely on them alone — not if we want to dream big and reach our potential. Unlike the Level 1 goals, where your progress towards your final dream may not be immediately obvious, the Level 2 goals should resemble a recognisable part of your bigger dream or goal. Using an analogy, say we have a 10,000-piece jigsaw puzzle. Level 1 goals would be the individual pieces, whereas Level 2 goals would form an entire section, revealing a main feature of the puzzle picture. Level 3 goals, which we will discuss next, would be akin to finishing an entire quarter of the puzzle, revealing the concept and details behind the picture, and making the finished design more obvious.

What would a Level 2 goal look like? When most of us make New Year resolutions, we say something like, "this year I'm going to lose 100lbs", or "by the end of the year I will buy a house". Level 2 goals would be the check-in period of, say, six months, where you would be aiming to have lost 50lbs or saved around 50% of the house deposit.

Take a moment to think deeply about each of your goals. Try not to make them so easy as to limit your progress, or too grand and unrealistic, which can lead to disappointment. Find a balance, stick to your plan with consistency and focus, and you will achieve it. Remember to set your Level 2 and 3 goals as challenging but achievable. Level 1 goals should always be easily attainable, but the longer-term goals will provide you with better performance and productivity when additional difficulty is added to them.[40]

Level 2 rewards — The fun part

These rewards are going to be something that you will look forward to for months, so try to put some thought into what they will be for you. Again, hold back on anything too extravagant just yet, but give yourself something worthwhile that feels like a well-deserved reward. It should feel special because that is exactly what it is. You will have worked incredibly hard to achieve your Level 2 goal, so be creative with it. Write it on the calendar or in your journal, post about it on social media, and tell your friends about it.

The reward itself will be personalised for you but could be anything from a day of shopping for yourself, a spa weekend, new video games or books, a weekend away with your family, going to a concert or live sports event, or anything that brings you joy and happiness. It doesn't have to cost money, as long as it is a noticeable time for you to reflect on your efforts and achievements from the previous months.

Level 3 goals

Level 3 goals, which are the (potentially multiple) year-long goals you set for yourself, will be tailored to your final dream. They will be natural checkpoints and powerful motivators because they allow clear progress to be demonstrated while still feeling less overwhelming than your final goal.

The Level 3 goals should be placed every 12 months, ideally at the end of the year, which creates a natural ending to your goals and targets. As with all your goals, be sure to set your Level 3 goals with a degree of difficulty. They should be challenging but achievable. The greater the challenge, the more likely you will push yourself to more extraordinary performance and overall achievement.[43] Make them ambitious because you have the time within those 12 months to adjust your schedule to ensure you achieve Level 3 goals.

Remember to tell friends and family when you achieve these goals. They are likely a part of your success and would love to be updated on your achievements. Allow those close to you to share in your joy and progress.

Level 3 rewards — The fun part

Well done. Even reading about achieving Level 3 goals shows your ambition and drive to achieve them, and because they are a big achievement, dedicate some time to think about what the rewards should be for you. The rewards do not need to be expensive. This will depend on your financial situation. The essential idea is to make the rewards different and unique to represent the prestige of achieving your Level 3 goal.

Take as long as you need to think about what these rewards will be for you. You will have been working towards them for 12 months, and so they should be worth the effort. Level 3 rewards could be anything from refurbishing a room in your house, a holiday, a new computer or TV, a family day out or weekend away, or simply a week off to share time with family and friends, taking the time to enjoy yourself.

As this type of goal is usually placed at the end of the year, the holiday season usually presents more time for friends and family. Use this to your advantage and plan your reward around the expected free time of this time of year. Take the time to relax, recharge and embrace and enjoy your rewards.

As with all of the rewards, make them personal and meaningful to you, and write them in your calendar or journal as a way to remind yourself of what is coming. Ultimately, the feeling within yourself when you understand the commitment and hard work that has taken you to this point is unmistakable, and one of the reasons why it is important to enjoy the process of productivity, and not simply the end result or dream. These goals and rewards along the way are just as important and valuable as any bigger dream you may have. So, please take the time to enjoy them!

9
MAKING A LASTING CHANGE

True change takes time, but it lasts forever.

Neuroplasticity

As you go through your journey and achieve your goals, you will begin to notice that tasks become more natural and 'second nature'. Over time, you will notice that there will be moments of 'when it feels right', meaning that you have a feeling of wanting to do a certain task in a certain moment above all others, and we will explore what to do in those times in the following pages. These moments happen because you learn and adapt to new behaviours and routines, and eventually, they become more familiar. This adaptation has a biological name — neuroplasticity.

When we learn and develop, we gain insight and experience that gives us a wider perspective for tackling new situations. Likewise, when we begin to make changes in our lives, such as adopting new methods to increase our productivity, we eventually become accustomed to and familiar with those changes. Something that may have initially appeared difficult now becomes easier, and from this point, we can search for further improvements, all of which lead us to our final goal. Gaining a new perspective on your schedule is just the beginning because we are looking for real long-term changes to your behaviour, and that occurs in the brain through neuroplasticity.

Neuroplasticity is the process by which your brain adapts to new environments and requirements to become better, and ready for the next time you encounter them. In fact, you may already be aware of this process, whether it is practicing a new skill and improving at it, learning to play a musical instrument, or recalling a memory. These are all examples of this brain adaptation.

In short, your brain is composed of around 86 billion neurons (the typical type of brain cell) and billions of other cells, such as microglial cells, oligodendrocytes, and astrocytes. Neurons communicate via chemical messengers released at their synapse in order to communicate with other neurons. This communication or 'connectivity' forms the basis of everything that your brain is able to accomplish.

When you learn something new, gain a new experience, adjust your schedule and behaviour, or form a new habit, this brain network changes; it adapts through neuroplasticity. New experiences also help us to improve and adapt. Even the brain waves and firing patterns of neurons change when we experience a new place that is unfamiliar.[45]

Research shows that in as little as a few weeks the brain can adapt by increasing white matter (the long axons of neurons where the messages travel along).[46] This process is accelerated when we expose ourselves to environments and workloads that engage us. These changes occur in many regions of the brain, but it is the frontal cortex, where our reasoning and decision–making occurs, which is particularly relevant for here.

A number of animal and human imaging studies have described how learning is closely linked with behavioural changes; that is, there are significant improvements to our lives when we push ourselves to improve and grow. Incredibly, these changes can occur at any age. Even in seniors, the brain continues to grow new neurons, particularly in our memory regions.

The most well–studied form of experience–based learning and neuroplasticity is called long–term potentiation, or LTP for short. LTP explains the neural mechanisms behind the adaptive behavioural changes that we make. This process occurs rapidly from around 30 minutes after a new experience and continues for hours or even days. Changes at the synapses of neurons, the place where the chemical messengers communicate, changes to the shape of brain cells, increased white matter, and many other adaptations can all occur under the process of LTP. The brain remodels itself through these changes to a more efficient way of operating. Memories can be stored and recalled when we need them, skills can be learned and developed, and our habits and behaviours can adapt to the new demands we make and become automatic.

Thousands of scientific studies have been dedicated to further understanding the cellular mechanisms that occur within the brain and the real–life applications of neuroplasticity.[46,47] By the end of the 20th century, neuroscience had already described the structural changes in the brains of professional musicians[48] to improve the communication between the two hemispheres of the brain; showed enlarged memory regions in London taxi drivers;[49] and demonstrated how a healthy lifestyle could grow new neurons.[50] The message is clear — your brain is always working to find ways to help you adapt and improve.

Experience–induced neuroplasticity will allow you to continue to develop and improve as your employ everything you are learning throughout the quantum method. Your goals might feel formidable at first, but stick with them, because things will get easier along the way.

"Any new experience can be scary, so remind yourself often how amazing you are for trying"

Experience counts

Essentially, experience provides meaningful changes in the brain. This experience counts for a lot when it comes to how we react to and view different situations. A research study looking at the brain using MRI scans (a way to see increased blood flow and therefore increased workload in specific areas) compared the brain between managers and entrepreneurs in decision–making tasks.[51] The study demonstrated how entrepreneurs could get the same result, but in less time, meaning their decision–making process was slightly more efficient. So, what does this tell us? The researchers suggested that entrepreneurs may have benefited from having more experience in handling decisions from a wider range of situations. All of those experiences change how a person tackles new problems and works under pressure. This will, of course, vary hugely depending on the individual, regardless of whether you fit into one of those professional categories, but the overall conclusion is clear. Prioritise putting yourself into new and unfamiliar situations so you can grow and improve, leading to better decision-making and overall performance and productivity throughout your journey.

Whenever you encounter situations pushing you out of your comfort zone, my message for you is this: **do it**! If you want to change your life, if you're going to achieve your dream, if you want to boost your productivity to the level required for you to be the best version of yourself, then throw yourself into situations that are both diverse and unfamiliar. You will learn to deal with the stress and anxiety of decision-making in unfamiliar and uncomfortable situations, which might just make the difference in reaching your goals.

Find those moments of joy

In some ways, this chapter is different to the rest of the book. It takes some of the structure out of the methods used and replaces it with a little more of you. We will discuss something that flies in the face of the quantum method, which should be utilised only when you feel is the right time. What does that mean — 'when it feels right'? There are going to be many occasions along your path to achieving your dreams where you simply do not want to do a task on your list, and you would rather do something else. This is perfectly normal, however productive you might be. Often, this 'something else' can be entirely different to your goals, and usually, these feelings need to be suppressed so we can concentrate on what needs to be completed. There are instances, however, where the message to change to a different task is your brain's way of trying to actually improve your productivity. In these instances, and even though you understand that the other tasks on your list should take priority, it could be that you would rather work on something that *excites* you more. This is one of the ways we can truly make a lasting change in routines and behaviours, because we can find joy within the smaller moments of our journey towards our dream. The neuroplasticity processes within our brain will help to cement those routines and subroutines that help us to be consistent in our schedule, but it is also the work that we enjoy and excites us, that will encourage us to keep working hard.

So, what should you do when you reach the dilemma of wanting to do something a little outside of your planned activity? First, you should take a step back and ask yourself one important question. Are you putting this task off because you don't want to do it, or are you embracing the excitement for another task? To be blunt, are you feeling lazy, or is it genuine enthusiasm for another task?

Again, I encourage you to be as honest as you can about this question. You don't need to say this question out loud, or even tell anyone else that you are thinking it, but you should be honest about your answer. If the reason for wanting to work on something different is because you don't really want to do it, then unfortunately, you will need to get it done, regardless of how mundane or inconvenient it is. We all go through this feeling, and there will be times when this feeling overtakes you, and no amount of motivation will be able to wash it away. In these moments, you will need to put your head down and carry on. However — and this is the good news — if there is another task that you genuinely want to start working on, even at the cost of sacrificing time for another task, then you should go ahead and do it! That's right, in these moments, feel free to brush aside the controlled and well–planned schedule, and go with your gut feeling.

Please do not mistake this concept as a way of discounting the methods discussed in this book. Instead, see it as a way to release the pressure valve on your productivity mindset, and unleash your natural motivation and passion for what you do. In moments like these, encourage the exciting feeling, and use it to remind yourself of why you are chasing that dream in the first place.

Want to write a book, but you are stuck on Chapter 3? Go ahead and write the ending, if that is what you really want to do. You can always go back to Chapter 3 later. Need to work on building a new client list? Take an old client out for dinner and enjoy your time with them, strengthening the relationship, and enjoying the benefits of your previous hard work. Need to keep working on your new home renovation but are more excited about what it will be like when it is finished? Buy a new piece of furniture before you really need it so you can enjoy the feeling of what it will be like when it is all completed.

It is easy to give in to the pleasant feeling and immediate satisfaction of postponing the hard work. Switching to a more exciting task can inspire you to work more vigorously because it offers you a taste of the finished dream. Use it to remind yourself of why you are working hard, and to enjoy some of the rewards from your success and hard work.

A quick note. By now, you will have identified dozens of potential tasks that will need to be completed to reach your goals. While you are creating your schedule, keep in mind that you can always highlight certain tasks that you know will be more enjoyable and engaging, or tasks that absolutely need to take priority. If you need to prioritise one or two activities above all others, you can single them out during this time. Suppose the time comes when your schedule doesn't go exactly to plan. In that case, you can choose to complete the high-priority objectives and remove the less important ones for a later time. Of course, the aim is to complete everything, but that may not always be realistic. In real-life scenarios, with busy home and work lives, unexpected curveballs appear, and it is well worth keeping these priorities and adaptions in mind. In addition, research tells us that prioritising a small number of tasks, ideally, those with the highest value to your goals, can lead to better performance, more time savings, and increased overall productivity.[52]

10
PUTTING IT INTO PRACTICE

This is my time. I have waited long enough. Now it is my time.

Hopefully, after reading through this book, you will have discovered techniques that can fit into your life and improve your productivity. You are now more aware of how you can find moments in your schedule for different types of tasks. You know how you can structure your week to get the most out of it, and what devices and methods you can use for each of them. We have covered a lot of ground, and so it is time to revisit what we have learned. We now need to simplify all of it, and make sure you are clear on the next steps to take after finishing this book.

What are the next steps?

1. Take your time to answer the questions from the introduction, which will give you insights into your preferred times for being productive throughout the week. It will also help you to understand how you prefer to learn new skills and keep your motivation high.

2. Create a hierarchy of your own needs that need to be met in order for you to achieve more. They don't need to be followed to the letter, but it is a good time to take a moment to understand what you want and need to be successful and improve your productivity. You can amend the hierarchy at any time, it doesn't need to be completed today.

3. List the tasks that you need to accomplish, at least for the next 2–4 weeks. Then group them by difficulty and time requirements into easy, medium, and hard categories.

4. Analyse your day by mapping out each day of the week, looking to cut any wasted time by at least 50%.

5. Identify when and how you could implement zonal working. This can be anywhere from one hour to all day, but 3–4 hours would be ideal. Think of them as accelerator programs for your productivity.

6. Learn how you can use technology and different resources throughout the day to ensure you can always fill your time with some aspect of to–do list.

7. Using your answers from the introduction, write out your Level 1, 2, and 3 goals.

8. For every level of goal you accomplish, you should reward yourself

for achieving them. Write out what those rewards will be so you can plan for them ahead of time.

9. Create a schedule for one week and make sure to ease into it. It will be difficult to change your schedule overnight, so instead look to implement 20% of your maximum effort each time. Improve your productivity by 20% increments every two weeks. Two weeks after you start, you can try increasing your effort by 40%. After that, increase it to 60%. Eventually, you will have changed your schedule and working habits to a level where you no longer feel you are adjusting anything, and you have reached a more automatic level.

A final word

The best results will come from the compounding effect of all of the chapters, meaning that combining the effect of all of them will produce the most improvement. However, this book also works if broken down and used in a way that works for you as an individual, and so it is possible to read through and utilise techniques from selected chapters, if you think that you would benefit from them. The martial artist Bruce Lee, famous for his fighting style and philosophical perspective on life, once said that we should "absorb what is useful, reject what is useless, and add what is specifically your own". Of course, I hope that nothing in this book would be considered useless, but I do fully embrace the concept that a person should be able to take useful advice and techniques and fit them into their own life. This is what I hope you take from this book.

When writing this book, that is what I always kept in mind. If someone is starting a new pursuit of their dream, then going through each chapter systematically will no doubt be the best option. But, if you are someone looking for quick tips to add into your already effective schedule, then I wanted the chapters to be short enough for you to do just that. If nothing else, I hope that it helps you to think about your time a little differently than you have done before. I want you to see how it can be used to help you, not hinder you. The idea that we don't have time to pursue our dreams is not true. Sure, finding those extra minutes will be tough at first, it definitely won't be easy, but the time is there, it really depends on how much that dream means to you.

To end with the famous quote from Walt Disney, "the way to get started is to quit talking and begin doing".

Bonus

There are some final tips and pointers that I wanted to share, that can add to your productivity. Some of these ideas, you may already be aware of. There is a reason for that. They work! Hopefully, by now you have realised that what may work for one person may not work for another, so take a look through this list and try them all out for yourself. Give them a real chance and decide for yourself if you notice any improvement. As with many of the concepts in this book, you may experience some initial resistance to a change in your normal schedule, but stick with them. You will need to get past the awkward stage where you instinctively want to revert back to old habits of what feels safe and familiar. That is a completely normal reaction, but the most productive people are able to quiet that instinctive voice and see things through with discipline and consistency. Some of the items on the list may not be possible for your current situation, for many reasons, and if that is the case, try out as many as you can. If you can think of replacements that would fit your personal situation better, feel free to use the items on the list as inspiration.

i. Wake up early. Between 4–6 am. You can get a lot done before sunrise and ensure productivity before any additional daily responsibilities appear.

ii. Don't use caffeine until after lunch time — you don't need it. Morning caffeine is more a habit than a necessity. This one will be tough at first. Use caffeine strategically, particularly in the afternoon.

iii. Pack your bag the night before, ready for the next day. This is a huge stress reducer for the morning rush.

iv. Prepare your clothes for the next day, including exercise clothes if needed. Taking a few minutes in the evening to do this will ensure you have more time in the morning and reduce the likelihood that you will forget an item you need for the day.

v. Spend 15 minutes at the end of each day cleaning, tidying, or organising your home or workspace. Naturally, this is the time when we really don't want to do it. However, waking up to a tidy home is a significant morale boost that can give you a great start to the day.

vi. Meal preparation can be a huge time saver if you get the chance to do it. Most meals we cook can be frozen and saved for months. Leftovers from a previous meal can be perfect for freezing for another time.

vii. Find one task that could be done by spending money and buying yourself some extra time. Of course, this will depend on individual circumstances. But, if a small cost each month can give you extra time and reduce your stressful day, it might be worth it.

viii. Keep a physical calendar on your wall and mark on important dates, deadlines and targets. Seeing something every day can act as a constant reminder for tasks. This can work towards your productivity, or it can highlight your rewards and enjoyable experiences to look forward to.

ix. Tell someone about your goal for that month and be held accountable for it. The additional feeling of not wanting to disappoint a friend can be a powerful motivator. You can also bask in the success of your achievements when you reach your goal.

x. Take the time to enjoy the small successes in your life. When you are fortunate enough to experience moments when things are going well for you, take some time to step back and embrace them. Keep these feelings in your mind, joy, and happiness for when you need them in the future.

xi. Although it may not be available for everyone, if you can find someone who has already achieved a similar goal that you are aiming for, then reaching out to them could prove hugely beneficial. Having a mentor has been shown to improve career trajectory in multiple industries. Using a mentor's experience to help you along your journey may be a valuable asset.[53]

xii. Exercise! People who exercise before they start work show a 15% increase in productivity for the day.[54] It doesn't need to be much, whatever you can fit in with your schedule.

Bonus: Remind yourself out loud how well you are doing and how great you are. Compliment yourself on your effort and acknowledge your progress and hard work, because it is hard. Even making a small change to improve your life and chase your dreams should be a cause for celebration. You are strong and exceptionally good at what you are dedicating your time to. Don't ever forget to tell yourself that as often as you need to hear it.

Now is the time to grab your dream with both hands and drag it into existence. It is your time. It is your time!

ABOUT THE AUTHOR

Mike Tranter is originally from the north of England, UK, where he began his scientific career studying how drugs work in our body. However, his true calling was for neuroscience. After a PhD in neuroscience, he spent years in research laboratories all over the world studying how the brain works, and how value-based decision making occurs, although it is in science communication, and opening up the world of neuroscience to everybody, that he enjoys the most.

EXPLORE OTHER BOOKS BY THE AUTHOR

An easy way to learn about the brain. The most interesting questions you have about the brain are finally answered.

◆ What is depression and does it change the brain?

◆ Do men and women have different brains?

◆ What are dreams and why do we have them?

This book makes the brain fun and easy to enjoy. Anyone who is curious about what really goes on in that mushy pink thing inside their head will enjoy this guide to the brain and neuroscience.

Join neuroscientist Mike Tranter PhD as he explains the brain in his unique and funny style. He answers questions that were submitted by the public, and the best part is, no scientific background is needed whatsoever.

Includes a chapter describing some of the strange mysteries about the brain, and a behind the scenes look at how cutting-edge neuroscience research will change the future.

Available in English, Spanish, French, German, and Italian

REFERENCES

1: Adams, A. (2014). www.gallup.com. https://news.gallup.com/poll/181289/majority-employees-not-engaged-despite-gains-2014.aspx

2: Settle, J. R., *et al.* (2017). Initial planning benefits complex prospective memory at a cost. *The Quarterly Journal of Experimental Psychology*; 70 (8).

3: Ryan, T. A., & Smith, P. C. (1954). *Principles of industrial psychology.* New York: Ronald Press.

4: Mace, C. A. (1935). Incentives: some experimental studies. *Reports from the Industrial Health Residency Board of London*; 72 (69).

5: Locke, E. A. (1968). Toward a theory of task motivation and incentives. *Organizational Behavior and Human Performance*; 3 (2).

6: Maslow, A. H. (1943). A theory of human motivation. *Psychological Review*; 50 (4).

7: Tay, L., & Diener, E. (2011). Needs and subjective well-being around the world. *Journal of Personality and Social Psychology;* 101 (2).

8: Apollo Technical. (2022). www.apollotechnical.com. https://www.apollotechnical.com/working-from-home-productivity-statistics/

9: von Stumm, S. (2016). Is day-to-day variability in cognitive function coupled with day-to-day variability in affect? *Intelligence*; (55).

10: Epton, T., *et al.* (2017). Unique effects of setting goals on behavior change: systematic review and meta-analysis. *Journal of Consulting and Clinical psychology*; 85 (12).

11: Iqbal, S. T., & Horvitz, E. (2007). Disruption and recovery of computing tasks: field study, analysis and directions. *Conference of Human Factors in Computing Systems.* Conference paper.

12: Putnam, A. L., *et al.* (2016). Optimizing learning in college: tips from cognitive psychology. *Perspectives on Psychological Science*; 11 (5).

13: Carpenter, S. K., *et al.* (2012). Using spacing to enhance diverse forms of learning: review of recent research and implications for instruction. *Educational Psychology Review*; 24 (3).

14: Aeon, B., *et al.* (2021). Does time management work? A meta-analysis. *PLoS*

ONE; 16 (1).

15: Malkoc, S. A., & Tonietto, G. (2019). Activity versus outcome maximization in time management. *Current Opinion in Psychology*; 29.

16: Gustavson, D. E., *et el.* (2015). Understanding the cognitive and genetic underpinnings of procrastination: evidence for shared genetic influences with goal management and executive function abilities. *Journal of Experimental Psychology General*; 144 (6).

17: Stöber, J., & Joormann, J. (2001). Worry, procrastination, and perfectionism: differentiating amount of worry, pathological worry, anxiety, and depression. *Cognitive Therapy and Research*; 25 (1).

18: Newport, F. (2015). www.gallup.com. https://news.gallup.com/poll/187982/americans-perceived-time-crunch-no-worse-past.aspx.

19: Ferrari, J. R., *et al.* (2007). Frequent behavioral delay tendencies by adults: International prevalence rates of chronic procrastination. *Journal of Cross-Cultural Psychology*; 38 (4).

20: Day, V., *et al.* (2000) Patterns of academic procrastination. *Journal of College Reading and Learning*; 30 (2).

21: Stephens, D. W., *et al.* (2004). Impulsiveness without discounting: the ecological rationality hypothesis. *Proceedings of the Royal Society of London*; 271 (1556).

22: Van Eerde, W. (2000). Procrastination: self-regulation in initiating aversive goals. *Applied Psychology*; 49 (3).

23: Krause, K., & Freund, A. M. (2014). How to beat procrastination - the role of goal focus. *European Psychologist*; 19 (2).

24: Steel, P., *et al.* (2018). Examining procrastination across multiple goal stages: a longitudinal study of temporal motivation theory. *Frontiers in Psychology*; 9 (327).

25: Rozental, A., *et al.* (2018). Targeting procrastination using psychological treatments: a systematic review and meta-analysis. *Frontiers in Psychology*; 9 (1588).

26: Zimmerman, B. J., & Martinez-Pons, M. (1992). Self-motivation for academic attainment: the role of self-efficacy beliefs and personal goal setting. *American Educational Research Journal*; 29 (3).

27: Atlassian. (2022). https://www.atlassian.com/time-wasting-at-work-infographic.

28: Nagy, A. (2020). www.businesswire.com. https://www.businesswire.com/news/home/20200519005295/en/

29: Locke, E. A., & Latham, G. P. (2002). Building a practically useful theory of goal setting and task motivation: a 35-year odyssey. *American Psychologist*; 57 (9).

30: Neubert, M. J. (1998). The value of feedback and goal setting over goal setting alone and potential moderators of this effect: a meta-analysis. *Human Performance;* 11 (4).

31: Weiner, B. (1992). *Human motivation: metaphors, theories, and research.* Sage Publications, Inc.

32: Korn, C. W., *et al.* (2016). Performance feedback processing is positively biased as predicted by attribution theory. *PLoS One*; 11 (2).

33: Belli, G. (2016). www.businessinsider.com. https://www.businessinsider.com/why-you-should-never-work-longer-than-90-minutes-at-a-time-2016-11.

34: Lord, R. G., *et al.* (2010). Self-regulation at work. *Annual Review of Psychology*; 61 (1).

35. Coffeng, J. K., *et al.* (2015). Physical activity and relaxation during and after work are independently associated with the need for recovery. *Journal of Physical Activity and Health*; 12 (1).

36: Berman, M. G., *et al.* (2008). The cognitive benefits of interacting with nature. *Psychological Science*; 19 (12).

37: Oppezzo, M., & Schwartz, D. L. (2014). Give your ideas some legs: the positive effect of walking on creative thinking. *Journal of Experimental Psychology: Learning, Memory, and Cognition*; 40 (4).

38: Baretta, D., *et al.* (2019). Implementation of the goal-setting components in popular physical activity apps: review and content analysis. *Digital Health*; 5.

39: McEwan, D., *et al.* (2016). The effectiveness of multi-component goal setting interventions for changing physical activity behaviour: a systematic review and meta-analysis. *Health Psychology Review*; 10 (1).

40: Ariely, D., & Wertenbroch, K. (2002). Procrastination, deadlines, and performance: self-control by precommitment. *Psychological Science*; 13 (3).

41: Kleingeld, A., *et al.* (2011). The effect of goal setting on group performance: a meta-analysis. *Journal of Applied Psychology*; 96 (6).

42: Milkman, K., & Brabow, K. (2020) www.scientificamerican.com. https://www.scientificamerican.com/article/why-feeling-close-to-the-finish-line-makes-you-push-harder/

43: Cheng, P. Y., & Chiou, W. B. (2010). Achievement, attributions, self-efficacy, and goal setting by accounting undergraduates. *Psychological Reports*; 106 (1).

44: Dai, H., & Li, C. (2019). How experiencing and anticipating temporal landmarks influence motivation. *Current Opinion in Psychology*; 26.

45: Wilson, J. (1978). Loss of hippocampal theta rhythm results in spatial memory deficit in the rat. *Science*; 201 (4351).

46: Hughes, E. G., *et al.* (2018). Myelin remodeling through experience-dependent oligodendrogenesis in the adult somatosensory cortex. *Nature Neuroscience*; 21.

47: Markham, J. A., & Greenough, W. T. (2004). Experience-driven brain plasticity: beyond the synapse. *Neuron Glia Biol*; 1 (4).

48: Schlaug, G., *et al.* (1995). Increased corpus callosum size in musicians. *Neuropsychologia*; 33 (8).

49: Maguire, E. A., *et al.* (2000). Navigation-related structural change in the hippocampi of taxi drivers. *PNAS*; 97 (8).

50: Liu, P. Z., Nusslock, R. (2018). Exercise-mediated neurogenesis in the hippocampus via BDNF. *Frontiers in Neuroscience*; 12 (52).

51: Laureiro-Martínez, D., *et al.* (2014). Frontopolar cortex and decision-making efficiency: comparing brain activity of experts with different professional background during an exploration-exploitation task. *Frontiers in Human Neuroscience*; 7 (927).

52: Fernbach, P. M., *et al.* (2015). Squeezed: coping with constraint through efficiency and prioritization. *Journal of Consumer Research*; 41.

53: Grecco, L. M. & Kraimer, M. L. (2020). Goal-setting in the career management process: an identity theory prospective. *Journal of Applied Psychology*; 105 (1).

54: von Thiele Schwarz, U., & Hanson, H. (2011). Employee self-rated productivity and objective organizational production levels. *Journal of Occupational and Environmental Medicine*; 53 (8).

This is your year!

Made in the USA
Coppell, TX
04 February 2023